Pollution Limits and Polluters' Efforts to Comply

Pollution Limits and Polluters' Efforts to Comply

The Role of Government Monitoring and Enforcement

Dietrich H. Earnhart and Robert L. Glicksman

STANFORD ECONOMICS AND FINANCE

AN IMPRINT OF STANFORD UNIVERSITY PRESS

STANFORD, CALIFORNIA

Stanford University Press
Stanford, California
©2011 by the Board of Trustees of the
Leland Stanford Junior University.
All rights reserved.

This book stems from a project funded under STAR Research Assistance
Agreement No. R-82882801-0 awarded by the U.S. Environmental
Protection Agency. The book has not been formally reviewed by the EPA.
The views expressed in this book are solely those of Dietrich Earnhart
and Robert Glicksman. The EPA does not endorse any products
or commercial services mentioned in this book.

Special discounts for bulk quantities of Stanford Economics and Finance
are available to corporations, professional associations, and other
organizations. For details and discount information, contact the special
sales department of Stanford University Press. Tel: (650) 736-1782,
Fax: (650) 736-1784
Printed in the United States of America on acid-free, archival-quality paper

Library of Congress Cataloging-in-Publication Data
Earnhart, Dietrich, author.
Pollution limits and polluters' efforts to comply : the role of government
monitoring and enforcement / Dietrich H. Earnhart and Robert L. Glicksman.
pages   cm
Includes bibliographical references and index.
ISBN 978-0-8047-6257-1 (cloth : alk. paper)—
ISBN 978-0-8047-6258-8 (pbk. : alk. paper)
1. Water—Pollution—Law and legislation—United States.   2. United States.
Federal Water Pollution Control Act.   3. Water—Pollution—Government
policy—United States.   4. Chemical industry—Waste disposal—United
States.   I. Glicksman, Robert L., author.   II. Title.
KF3790.E27   2011
344.7304'6343—dc22   2011003139

Typeset by Westchester Book Services in 10/12 Sabon.

Contents

List of Tables and Figures

TABLES

FIGURES

Acknowledgments

The authors wish to thank several people for their assistance and guidance. First, we are grateful for the participation of Donald Haider-Markel and Tatsui Ebihara, who represent the other members of the research team funded by the EPA STAR grant. We regret that neither of them was able to participate in this book project. Second, we thank our dutiful research assistants: Dylan Rassier, J. Mark Leonard, Trisha Shrum, and Julia Brandes. Third, we thank Steven Maynard-Moody and Susan Mercer at the University of Kansas Institute for Policy and Social Research for their support—both financial and moral—and their insight, patience, and understanding. Fourth, we thank Steve Rubin at the Environmental Protection Agency for providing the many, many extracts of data from the Permit Compliance System database and explaining the data elements held within. Fifth, we thank the deans during whose tenure this book was written, including Gail Agrawal at the University of Kansas School of Law and Fred Lawrence at the George Washington University Law School, for affording us time and encouragement to complete this project. Sixth, we thank several colleagues at other research institutions and the EPA, such as Wayne Gray and Jon Silberman, for providing insight on our research.

Dietrich Earnhart wishes to dedicate this book to his loving wife, Sharon Ashworth, who encouraged him to pursue the aforementioned EPA STAR grant in 2000 even though his plate was full and overflowing. Robert Glicksman wishes to dedicate this book to his wife, Emily Glicksman, for her love and support in all things, great and small.

Pollution Limits and Polluters' Efforts to Comply

CHAPTER ONE

Introduction

1.1. THE PROMISE AND PITFALLS OF
THE CLEAN WATER ACT

In 1972, Congress adopted one of the nation's landmark environmental laws, the Federal Water Pollution Control Act Amendments, now known as the Clean Water Act. Passage was prompted by concern over the egregious state of the quality of the nation's coastal waters, rivers, lakes, and streams, some of which were so contaminated by industrial chemicals that they caught fire. Others were befouled with oil from events such as the Santa Barbara, California, oil spill in January 1969, which produced images on television news broadcasts of oil slicks and oilsoaked birds and marine mammals. The new law sought to "restore and maintain the chemical, physical, and biological integrity of the Nation's waters" by establishing a goal of eliminating the discharge of pollutants into navigable waters by 1985.

That lofty goal was obviously not achieved by the target date and is still little more than a distant and perhaps impossible dream. There is little question, however, that over the nearly four decades since the enactment of the Clean Water Act, enormous progress has been made in cleaning up the nation's waters. The quality of major rivers such as the Hudson and the Potomac has improved sufficiently to allow recreational uses that most would not have dared to pursue in the late 1960s and early 1970s. Few would dispute the notion that the Clean Water Act has been an enormously successful pollution abatement initiative.

Yet, as the first decade of the twenty-first century approached an end, and with the fortieth anniversary of the Clean Water Act on the horizon, trouble was brewing. From many quarters, disturbing reports of failures to enforce the law streamed in. In a lead editorial published late in 2009, the *New York Times* wrote that "even its staunchest allies agree than the act has grown old and fallen well short of its goals, crippled by uneven and sometimes nonexistent enforcement by state and federal agencies" (*New York Times*, 2009).

The newspaper's concern was triggered by investigative reports from the paper's own reporters. In an exposé published in September 2009, the *Times* reported that more than five hundred thousand known violations of the Clean Water Act had occurred during the period 2004–2007 by more than twenty-three thousand facilities, according to records submitted by the polluters themselves. Those figures likely underestimated the scale of the problem because some facilities engaged in illegal discharges fail to inform the government of these violations. According to environmental groups, the number of Clean Water Act violations had increased significantly in recent years. The *Times* reported that the number of facilities violating the Clean Water Act increased by more than 16 percent between 2004 and 2007. Worse, about 60 percent of the violations qualified as "significant," a term used to identify violations posing the highest public health or environmental risks (Duhigg, 2009).

The occurrence of frequent violations was bad enough news in itself, but that was only part of the disturbing story. According to the *Times*, fewer than 3 percent of Clean Water Act violations during the period investigated by the paper resulted in fines or other significant punishments by state officials; moreover, the federal Environmental Protection Agency (EPA) did little to press the states to take their enforcement responsibilities more seriously or to step into the breach created by inadequate state enforcement. State officials blamed the absence of vigorous enforcement despite high rates of regulatory noncompliance on increased workloads and dwindling resources: state enforcement budgets remained essentially flat when adjusted for inflation even though the number of regulated facilities more than doubled in the previous ten years. The *Times* concluded that state regulators often lacked the ability or training to levy fines large enough to deter polluters. Even though EPA acknowledged the problem, it hesitated to pressure the states to do better, partly because of its reluctance to risk putting stress on its relationships with state enforcement officials and partly because it lacked a consistent national oversight strategy (Duhigg, 2009).

Other contemporaneous accounts were consistent with the *Times*' findings. In testimony given before Congress the month after the *Times* story appeared, the U.S. Government Accountability Office (GAO) noted that while overall funding for carrying out enforcement activities in EPA's regional offices and in states authorized to issue and enforce Clean Water Act permits had increased from fiscal year 1997 through fiscal year 2006, those increases failed to keep pace with inflation and the agencies' increased enforcement responsibilities. More specifically, funding to EPA regional offices increased from $288 million in fiscal year 1997 to $322 million in fiscal year 2006 but declined in real terms by 8 percent over that period. The decline in funding was reflected in a decrease in full-time employees in many of EPA's regional offices. Essentially, both EPA and state officials felt overwhelmed by increased responsibilities and declining resources to meet them. The GAO concluded that "our work over the past 9 years has shown that the Clean Water Act has significantly increased EPA's and the states' enforcement responsibilities, available resources have not kept pace with these increased needs, and actions are needed to further strengthen the enforcement program" (GAO, 2009, p. 14). Among the specific consequences of the collapse of effective enforcement described by the GAO was a decline in the value of injunctive relief, which for purposes of its report the GAO defined as the monetary value of future investments necessary for an alleged violator to come into compliance. Reviewing EPA's assessed penalties from fiscal year 1998 to fiscal year 2007, the GAO found that total inflation-adjusted penalties declined from approximately $240.6 million in fiscal year 1998 to only $137.7 million in 2007 (GAO, 2009).

In the face of these troubling depictions of the state of Clean Water Act compliance and enforcement, EPA released a Clean Water Act Enforcement Action Plan in October 2009. The plan concedes forthrightly that Clean Water Act "violations are still too widespread, and enforcement too uneven" (EPA, 2009, p. i). It finds that "many of the nation's waters are not meeting water quality standards, and the threat to drinking water sources is growing" (p. i). Although EPA found that some states had strong water-quality protection and enforcement programs, compliance and enforcement vigor were uneven. Like the *Times*, EPA found unacceptably high rates of significant noncompliance—about 24 percent among the nation's largest direct-discharge facilities. EPA data reveal even higher rates (about 45 percent) of serious noncompliance (which the GAO equated with EPA's concept of "significant noncompliance") at smaller facilities that submit discharge-monitoring reports. Yet, according to EPA, the states reported taking enforcement action against fewer than 6 percent of these smaller facilities (EPA, 2009).

The absence of consistent enforcement by EPA and the states created an unlevel playing field for businesses complying with the law and for citizens seeking protection against the health and environmental risks posed by unlawful water-pollution discharges. The action plan pronounced that effective enforcement programs create incentives for compliance by penal-izing those who do not follow the law. They establish a level playing field between those members of the regulated community who comply and those who do not. Enforcement ensures fair treatment—companies that compete against each other should not face wide disparities in treatment across the country, such as mandatory minimum penalties for a violation in one state and no enforcement in another. Ultimately, enforcement is critical to ensure that the public receives the services and protections promised by our laws. Unfortunately, data shows us that we are not getting the compliance envi-sioned by our laws to protect clean water (EPA, 2009, p. 6).

EPA's assessment of the state of Clean Water Act compliance and en-forcement led it to conclude that new approaches were needed to revamp its enforcement program so that EPA and the states would focus their enforcement efforts on the Clean Water Act violations that posed the big-gest threats to water quality and public health, including a reinvigoration of both civil and criminal enforcement against traditional end-of-pipe pollution. Testifying before Congress at the same hearings at which the GAO appeared, EPA administrator Lisa Jackson announced the formula-tion of the new action plan, proclaiming that "the time is long overdue for E.P.A. to re-examine its approach to Clean Water Act enforcement" and that EPA's goal was to "develop more innovative approaches to target enforcement to the most serious violations and the most significant sources" (Duhigg, 2009).

I.2. CONTRIBUTION OF THE PRESENT STUDY

This book seeks to provide insights into the impacts of Clean Water Act en-forcement on both performance and behavior by facilities regulated under the statute. In doing so, its goals include providing information to EPA and the states responsible for implementing and enforcing the Clean Water Act. We anticipate that the information may assist them in fashioning the kind of innovative and effective enforcement programs that EPA administrator Jackson has identified as necessary for providing the fair treatment of regulated facilities, effective public health and environmental protection, and achievement of the goals and promise that the Clean Water Act staked out in 1972.

The analysis in this book is based on information on Clean Water Act enforcement that relates to the same period of time analyzed by the GAO testimony and EPA Clean Water Act Enforcement Plan released in October 2009. In particular, the data we analyze here measure facility discharges and enforcement actions taken during the period 1999 to 2003. The study focuses on discharges by, enforcement actions taken against, and inspections conducted of discharging facilities in the chemical industry, one of the most significant industries regulated by the Clean Water Act and one that on occasion EPA has designated as a priority industrial sector. We believe that these aspects of the chemical industry make it a valuable focus of an empirical analysis of the relationships among the imposition and enforcement of discharge limits under the Clean Water Act and environmental behavior and performance, notwithstanding the possibility that operation and compliance practices differ among the various industries regulated under the act.[1]

The book examines several broad research questions. These questions ask what the variations are in the discharge limits that apply to discharging facilities with wastewater permits; what actions discharging facilities are taking to comply with their discharge limits; what outcomes (in terms of discharges and compliance) result from various forms of environmental behavior; and what steps federal and state regulators are taking to induce compliance with discharge limits. We seek to ascertain whether different forms of enforcement actions and inspections help to induce better environmental behavior or better environmental performance by regulated facilities. We also assess how discharge limits affect environmental behavior and performance and how environmental behavior affects environmental performance.

Our goal in analyzing these issues with respect to facilities in the chemical industry is to provide information that may be useful to environmental policy makers in both the federal and state governments in designing regulatory and enforcement programs that induce improvements in environmental performance and desirable changes in behavior by regulated facilities. By studying the impact of past regulatory activity—crafting discharge limits for polluting facilities in the chemical industry and pursuing enforcement actions against facilities alleged to have violated their regulatory obligations under the Clean Water Act—we should be able to provide useful information so that policy makers may be able to maximize the degree to which regulatory expenditures create the greatest degree of improvements in environmental compliance.

Although other empirical studies cited throughout this book assess the impacts of certain regulatory decisions on environmental performance and behavior, we are not aware of any empirical study that engages in the kind of sweeping evaluation of a broad range of regulatory actions on an important polluting industry that we undertake in this book. Further, the fact that the data surveyed and analyzed here derive from the very period that both EPA and the GAO have pointed to as evidence of the failure of past federal and state enforcement approaches affords us a unique opportunity to determine the extent to which particular kinds of regulatory and enforcement efforts have succeeded or contributed to past enforcement failures.

1.3. PROVISIONS OF THE CLEAN WATER ACT

The Clean Water Act announces as its ultimate goal the elimination of all discharges of pollution to the nation's waterways and as an interim goal the achievement of fishable, swimmable waters. The principal legal tool for achieving these goals is a provision that makes it unlawful to discharge pollutants into waters of the United States without a permit. The Clean Water Act creates two permit programs, only one of which is relevant to this book. That program is the National Pollutant Discharge Elimination System, or NPDES, permit program. This program is administered by states that EPA has authorized to issue individual discharge permits for point sources of pollution, or by EPA itself in states that have not been so authorized. Point sources are those that discharge pollutants through discrete conveyances, such as pipes, rather than through diffuse runoff. The second permit program is called the Section 404, or dredge-and-fill, permit program, which deals with the discharge of dredged or fill material, primarily to wetlands. Industrial pollutant discharges, such as the discharges of total suspended solids and biological-oxygen-demanding material by chemical industry facilities, which are the focus of this book, do not implicate the Section 404 program. Therefore, we do not discuss this second program further.

The Clean Water Act requires that a permit impose discharge limits on regulated sources. These limits restrict the quantity or concentration of pollutants that sources may discharge into the nation's waterways. EPA issues regulations that contain effluent limitation guidelines that apply to entire industrial categories of sources. These limitations are based on the degree of pollution reduction that EPA determines is achievable through the use of technology that is available to the industry concerned. Permit-issuing agencies use the regulatory limitations as the starting point for determining the discharge limits to impose on individual point sources applying for a wastewater discharge permit. Those discharge limits may differ from the regulatory limitations for any number of reasons (which Chapter 4 describes in more detail). The Clean Water Act allows the states to impose discharge limits that are more stringent than those demanded by EPA. If a state environmental agency decides to exercise that authority, the discharge limit it imposes on a source in its NPDES permit may be more stringent than the limitations adopted in EPA's effluent limitation guidelines for the relevant industry.

Another reason that the individual discharge limits for a particular point source may differ from the regulatory limitations relates to water-quality standards adopted by the states. The Clean Water Act requires that every state adopt and periodically revise water-quality standards designed to protect the public health and welfare, enhance water quality, and serve the Clean Water Act's purposes. If a state determines that the quality of a particular body of water, such as a stream or lake, is not adequate to meet the applicable water-quality standard, it is responsible for devising strategies for reducing pollutant concentrations in that waterway to the extent needed to bring the waterway into compliance with the water-quality standard. One way to do so is to impose discharge limits on point sources in NPDES permits that are more stringent than those that appear in EPA's effluent limitation guidelines.

The Clean Water Act also establishes an extensive enforcement program. The statute authorizes EPA to establish reporting requirements on point-source discharges, inspect regulated facilities, and initiate appropriate enforcement action against sources alleged to be discharging without a permit or to be in violation of their permits or other regulatory obligations. Although the statute allows the federal government to bring criminal charges against violators, the focus of this book is on the Clean Water Act's civil enforcement provisions. The statute authorizes EPA to pursue informal enforcement actions, impose administrative penalties, and issue civil administrative orders that mandate actions needed to bring regulated sources into compliance. Alternatively, EPA may bring suit in federal court seeking the same kinds of relief. In some cases, the courts also have been willing to require violators to pursue supplemental environmental projects to mitigate or offset the environmental harms the actions have caused. Although both proceedings in court to impose monetary penalties or injunctive relief and administrative proceedings seeking the same remedies are properly characterized as civil (as opposed to criminal) proceedings, we use the term *civil* in this book exclusively in connection with judicial proceedings. We refer to enforcement actions resolved by EPA or state agency officials rather than by judges as administrative proceedings.

The Clean Water Act also envisions enforcement by the states. Indeed, one of the conditions that a state must meet before it is eligible to jointly administer the NPDES permit program is a demonstration to EPA that it has adequate legal authority and resources to enforce the law. The manner in which states are allowed to enforce the Clean Water Act permits they issue differs in accordance with the laws of each state. Generally, however, the states have the authority to demand the same kinds of reports that EPA requires of regulated sources, inspect those sources, and

impose (or seek judicial imposition of) civil penalties and injunctive orders on violators. EPA retains the power to enforce the Clean Water Act
even in the states that it has authorized to administer and enforce the
NPDES permit program. This book examines both permits issued and
inspections conducted by state agencies but not enforcement actions
taken by state agencies.

Chapters 4 and 7 describe more fully the provisions of the Clean Water
Act summarized here.

1.4. HISTORICAL DATA

This last section provides data on the amounts of pollutants discharged
into the nation's waters, the compliance status of regulated facilities, and
the number of enforcement actions taken and inspections made by EPA
and the states during the period covered by our study (1999–2003). These
figures provide background for the analysis of the questions we pose
throughout the book concerning discharge limits, environmental behavior, environmental performance, government interventions, and the relationships among these variables.

1.4.1. Wastewater Discharges

The chemical industry discharges significant amounts of pollutants
into the nation's waters, making its performance and behavior important to an assessment of the impact of Clean Water Act enforcement. As
Table 1.1 illustrates, in the years covered by our study (1999–2003), the
chemical and allied products sector discharged between 44.5 and 77.1 million pounds of pollutants into the nation's surface waters. The amount
discharged declined each year (data for 2002 are not available), as did
chemical industry discharges as a percentage of the total amount discharged
by all industrial sectors. Nevertheless, the percentage of the total amount
discharged by all industrial sectors that was discharged by the chemical
and allied product sector remained substantial—between 20 and 29.8
percent.

1.4.2. Government Interventions: Inspections
and Enforcement Actions

Chapter 7 addresses the degree to which EPA and the states engaged in
inspections of facilities in the chemical industry with NPDES permits and
took enforcement actions against those facilities. Chapters 8 and 9 ana-

TABLE I.I
Surface-water discharges, 1999–2003

Year	Surface-Water Discharges— Total (millions of pounds)	Surface-Water Discharges—Chemical and Allied Products Sector (millions of pounds)	Percentage of Total Discharges by Chemical and Allied Products Sector
1999	258.9	77.1	29.8
2000	260.9	68.7	26.3
2001	220.8	57.6	26.1
2002	n.d.	n.d.	n.d.
2003	222.6	44.5	20.0

SOURCES: *Statistical Abstract of the United States* 2001, Table 363; 2002, Table 355; 2004–2005, Table 367; 2006, Table 367. Washington, DC: U. S. Census Bureau.

NOTES: Based on reports filed under Section 313 of the Emergency Planning and Community Right-to-Know Act, 42 United States Code § 11023 (2006), facilities within Standard Industrial Classification (SIC) Codes 20 through 39 with 10 or more full-time employees

lyze the impact of these inspections and enforcement actions (which we refer to collectively as "government interventions") on environmental behavior and environmental performance in this industrial sector. To provide a backdrop for this analysis, we depict here the degree to which EPA engaged in government interventions across all industrial sectors during the period covered by our study. Table 1.2 provides the dollar values of different forms of formal enforcement action, both civil and criminal (although this book focuses exclusively on civil enforcement). The data indicate that the dollar value of civil judicial penalties assessed across all industrial sectors varies significantly during this period, ranging from about $7 million in 1999 to about $59 million in 2003. The range of dollar values for administrative civil penalties assessed during this period is much smaller (between about $5 and $7 million), and the dollar amounts assessed are significantly less than the amounts assessed in judicial proceedings in every year except 1999. The dollar value of combined injunctive relief provided in both civil judicial and administrative proceedings far outstrips the value of either judicial or administrative civil penalties assessed in every year for which figures are available, although the dollar value of injunctive relief varies widely over the time period covered by our study. For the years for which data are available, the dollar value of judicial injunctive relief outstrips the value of administrative injunctive relief and is far greater than the value of either judicially or administratively assessed civil penalties for the same year. The dollar value of supplemental environmental projects (SEPs) pales in comparison with the dollar value of either judicial or

TABLE 1.2

Clean Water Act enforcement activity, fiscal years 1999–2003

Fiscal Year	Criminal Penalties Assessed	Civil Judicial Penalties Assessed	Administrative Penalties Assessed	Dollar Value of Injunctive Relief	Dollar Value of Judicial Injunctive Relief	Dollar Value of Administrative Injunctive Relief	Dollar Value of SEPs
1999	26,344,626	7,416,728	5,200,575	577,486,331	n.d.	n.d.	8,620,321
2000	49,901,801	21,579,394	5,403,201	156,813,072	n.d.	n.d.	10,857,998
2001	46,471,389	17,984,220	5,554,194[a]	168,587,320[b]	118,757,092	60,925,238	3,390,528
2002	35,884,399	16,951,009[c]	4,940,169	2,305,638,458	1,788,732,451	813,784,601	13,078,678
2003	n.d.	59,829,967	6,816.642	n.d.	n.d.	n.d.	9,344,514

SOURCES: EPA, FY 1996 Enforcement and Compliance Assurance Accomplishments Report, p. A-2, available at http://www.epa.gov/compliance/resources/reports/accomplishments/oeca/fy96accomplishment.pdf; EPA, Enforcement and Compliance Assurance Accomplishments Report FY 1997, p. A-2, available at http://www.epa.gov/compliance/resources/reports/accomplishments/oeca/fy97accomplishment.pdf; EPA, Enforcement and Compliance Assurance: FY98 Accomplishments Report 93, available at http://www.epa.gov/compliance/resources/reports/accomplishments/oeca/fy98accomplishment.pdf; EPA, Annual Report on Enforcement and Compliance Assurance Accomplishments in 1999, p. B-1; EPA, Protecting the Public and the Environment Through Innovative Approaches: Fiscal Year 2001 Enforcement and Compliance Assurance Accomplishments Report 72, EPA300-R-02-010; EPA, Environmental Results Through Smart Enforcement: Fiscal Year 2002 Enforcement and Compliance Assurance Accomplishments Report 59, EPA300-R-03-002; EPA, "FY 2000–FY 2008: Administrative and Civil Judicial Penalties Assessed," available at http://www.epa.gov/compliance/resources/reports/nets/nets-f3-adminandjudpen.pdf; EPA, "FY 1999–FY 2008: Supplemental Environmental Projects (SEPs)," available at http://www.epa.gov/compliance/resources/reports/nets/nets-f4-seps.pdf.

NOTES: SEPs, supplemental environmental projects; n.d., no data available.

[a]This figure appears as $5,353,442 in "Dollar Value of FY 2001 EPA Enforcement Actions (by Statute)," available at http://www.epa.gov/compliance/resources/reports/endofyear/eoy2001/eoyfy2001$law-rpt.pdf, reported as of January 23, 2002.

[b]This figure appears as $208,587,320 in FY 2001 Measures of Success Management Report, p. ES-3.

[c]This figure appears as $8,706,339 in "Dollar Value of FY 2002 EPA Enforcement Actions (by Statute)," available at http://www.epa.gov/compliance/resources/reports/endofyear/eoy2002/mosfy2002dveabystatute.pdf, reported as of December 9, 2002.

administrative injunctive relief but is fairly comparable to the value of civil judicial penalties assessed for most of the years covered by our study.

Table 1.3 provides information about the number of inspections (by both EPA and the states) of both major and minor facilities with NPDES permits and various kinds of informal and formal enforcement actions taken by EPA during the period covered by our study. Although data are not available for every year we studied, the figures show that the states were far more active in inspecting facilities with NPDES permits than EPA was. This result is not surprising given that some forty-five states have received permission from EPA to administer the NPDES permit program (that is, they have primacy in NPDES permit enforcement, with EPA retaining a backup role). The data show a marked and consistent decline in the number of informal actions and notices of violation issued over the period we studied. The number of administrative compliance orders issued and administrative penalty order complaints remains fairly constant over time, by contrast. Finally, EPA entered into relatively few civil judicial settlements during the entire five-year period we studied.

Clearly, we could provide much more information on the implementation of the NPDES program. However, we must limit the length of this book. Fortunately, we are able to offer additional information in the form of tables on the Web site of the Stanford University Press. We refer to these additional tables in certain points of the book. For example, the Web site offers a table that provides data concerning informal and formal enforcement actions taken by EPA's regional offices during fiscal year 2001. Perhaps the starkest figure to emerge from these data is the wide divergence among the regions. Regions 1 and 10 were relatively inactive during the period we studied, whereas Regions 2 and 6 were active in pursuing both civil penalties and SEPs.

1.5. SUMMARY

The Clean Water Act is responsible for significant improvement in the quality of the nation's surface resources. Despite progress toward the statutory goal of achieving water quality suitable for fishing and swimming across the nation, recent reports indicate that violations of the statute and of the regulations and permits issued under it continue to occur with troublesome frequency. In 2009, EPA embarked on a reexamination of federal and state approaches to enforcement to help adjust its enforcement program to achieve more consistent and effective enforcement. This book provides an empirical investigation of the compliance status of point sources in the

TABLE 1.3

National Pollutant Discharge Elimination System (NPDES) enforcement activity, fiscal years 1999–2003

Fiscal Year	EPA Regional Inspections		State Inspections		Informal Actions/ Notices of Violation	EPA Administrative Compliance Orders Issued	EPA Administrative Penalty Order Complaints	EPA Administrative Penalty Settlements	EPA Civil Judicial Settlements
	NPDES Majors	NPDES Minors	NPDES Majors	NPDES Minors					
1999	965	949	n.d.	n.d.	2080	549	136	186	16
2000	1141	640	9830	29,585	1062	596	295	291	12
2001	876	758	8907	27,064	565	367	192	210	24
2002	1044	871	n.d.	n.d.	455	490	155	131	33
2003	1221	595	n.d.	n.d.	n.d.	558	208	172	19

SOURCES: EPA, *Annual Report on Enforcement and Compliance Assurance Accomplishments in 1999*, pp. B-2–B-5; EPA, *Protecting the Public and the Environment Through Innovative Approaches, Fiscal Year 2001, Enforcement and Compliance Assurance Accomplishments*, pp. 66–69, 100; EPA, *FY 1994–FY 2008 Federal Inspection/ Evaluation Trends*, available at http://www.epa.gov/compliance/resources/reports/nets/nets-g2-inspectionslongterm.pdf; EPA, *National Enforcement Trends*, available at http://www.epa.gov/compliance/resources/reports/nets/nets-e1-apocomplaints.pdf, pp. E-1b, E-4b, E-5b, E-7b; EPA, *FY 2001 Measures of Success Management Report*, Washington, D.C. p. C-3.

NOTE: n.d., no data available.

chemical industry and of federal and state enforcement initiatives under the Clean Water Act during the period 1999–2003. This analysis should be useful to environmental policy makers at both the federal and state levels intent on creating and implementing effective enforcement programs that create incentives for regulated facilities to comply with the law and, ultimately, that translate into improved water quality with reduced risks to the public health and the environment. Among the topics the book addresses are the relationships between environmental behavior and environmental performance, between monitoring and enforcement and environmental behavior and performance, and between discharge limits of varying stringency and environmental behavior and performance.

1.6. ORGANIZATION OF THE BOOK

This book consists of ten chapters, including this introductory one. Chapter 2 explains how we selected the sample of regulated facilities in the chemical industry that provides the basis for our empirical investigation of regulatory limits, environmental behavior, environmental performance, and government interventions. It describes why the chosen sample facilitates our understanding of these topics and why the chemical and allied products sector serves as an excellent vehicle for examining corporate environmental performance under the Clean Water Act. The chapter also describes our survey of the identified facilities, which we designed and implemented in order to gather information on facilities' environmental management practices and perspectives on the effectiveness of government interventions.

Chapter 3 summarizes the research questions we explore in this book and places those questions in the context of previous empirical studies of discharge limits, environmental behavior, and environmental performance. Broadly, the book addresses four research questions. First, What is the variation in Clean Water Act discharge limits imposed on discharging facilities? Second, What actions (what environmental behavior) are discharging facilities taking to comply with those limits)? Third, What are the outcomes (what is the environmental performance)—in terms of discharges and compliance—of those actions? Fourth, What are regulators doing to induce compliance with the imposed discharge limits through inspections and enforcement actions (that is, government interventions)?

Chapter 4 addresses the imposition of effluent limits. It describes the structure of the Clean Water Act, the roles that EPA and the states play under the act in setting and enforcing discharge limits for particular discharging facilities, and the operation of the NPDES permit program in imposing and enforcing these limits. We assess in this chapter whether a relationship

exists between limits on the two pollutants we choose to study—total
suspended solids and biological oxygen demanding material—and limits
imposed on other pollutants discharged by the chemical manufacturing
industry. The chapter lastly assesses the presence and stringency of limits
across regions, across industrial subsectors, and over time.

Chapter 5 addresses environmental behavior. It centers on the relationship between discharge limits and environmental behavior and tests the
hypothesis that tighter limits are effective at prompting better environmental behavior. The analysis involves examination of the impact of limits on
various forms of behavior, such as the presence or absence of compliance
with industry-generated international standards for environmental management systems and the number of environmental employees working
at a particular facility.

The focus of Chapter 6 is environmental performance. The central inquiry is how discharge outcomes—measured in terms of absolute discharges
and the ratio of absolute discharges to permitted discharges (the discharge
ratio)—change as discharge limits vary. We test two hypotheses: (1) environmental agencies can effectively reduce absolute discharges by tightening
limits, and (2) tighter limits increase discharge ratios, indicating that limits
constrain discharging facilities.

In Chapter 7, we deal with regulatory efforts to induce compliance with
discharge limits. We assess what government interventions were taken by
EPA and the states against chemical manufacturing facilities during our
sample period. This chapter describes the legal authority of EPA and the
states to enforce discharge limits under the Clean Water Act and the different types of inspections and enforcement actions available to regulators.
The chapter explores government interventions taken against all major facilities in the chemical industry regulated under the Clean Water Act during
our sample period and against only the facilities that participated in our
survey. We analyze how enforcement activity varies across EPA regions,
time, subsectors within the chemical manufacturing sector, and categories
of facilities based on size.

Chapter 8 assesses the effects of government interventions on environmental behavior. After describing the difference between specific and
general deterrence, we test the hypothesis that greater deterrence prompts
better environmental behavior by examining the effects of government
intervention-based deterrence on the various measures of environmental
behavior, identified in Chapter 5.

Similarly, Chapter 9 deals with the effects of government interventions
on environmental performance. This chapter first describes the perceptions
of our survey respondents concerning the impact of interventions on environmental performance. Then the chapter tests through multivariate

regression analysis the hypothesis that greater deterrence prompts better environmental performance.

Chapter 10 provides a summary of the conclusions we reach based on our empirical analysis, the policy implications of those conclusions, and our suggestions for future research to facilitate a better understanding of pollution limits, regulated facilities' efforts to comply with those limits, government efforts to induce compliance through inspections and enforcement actions, and the effectiveness of those interventions in lowering discharges and improving compliance.

CHAPTER TWO

Scope of Analysis:
Sample of Regulated Facilities

2.1. SELECTION OF SAMPLE

In order to explore empirically the topics of regulatory limits, environmental behavior, environmental performance, and government interventions, we select a particular sample of regulated entities. Specifically, we select chemical manufacturing facilities that were permitted within the National Pollutant Discharge Elimination System (NPDES)—by either the Environmental Protection Agency (EPA) or the relevant state environmental protection agency—to discharge wastewater into surface-water bodies (hereafter "permitted wastewater-discharging facilities"). These facilities must have been operating in 2001 and 2002. We surveyed these facilities in 2002 and 2003. Based on the structure of our survey, the full sample period expands to January 1999 through March 2003. As described in Chapter 4, we assess all of the wastewater pollutants regulated across the full set of facilities. However, we focus our analysis on two specific pollutants: total suspended solids (TSS) and biological oxygen demand (BOD).

2.2. CHOSEN SAMPLE

This sample is worthy of exploration for a variety of reasons. First, the chosen medium of wastewater discharges facilitates our understanding immensely. Regulators systematically record wastewater discharge limits and actual discharges, procedures they do not take with other media. While EPA maintains records on compliance with regulations limiting air pollution, hazardous waste, and toxic waste, along with other pollution-related activities, these data provide little information about the imposed regulatory limits on specific regulated entities, undermining our analytical ability to assess the degree of compliance, especially overcompliance.

Second, the chosen sector of chemical and allied products serves as an excellent vehicle for examining corporate environmental performance. EPA has demonstrated a strong interest in this sector, as evidenced by its study—undertaken jointly with the Chemical Manufacturing Association (CMA)—on the root causes of noncompliance in this sector (EPA, 1999) and EPA's study on the compliance history of this sector, *Chemical Industry National Environmental Baseline Report 1990–1994* (EPA, 1997). Consistent with this interest, two of the chemical industrial subsectors—industrial organic chemicals (Standard Industrial Classification Code 2869) and chemicals and chemical preparations (Standard Industrial Classification Code 2899)—were regarded by EPA as priority industrial sectors during a portion of the sample period. Similarly, this sector is expected to display a substantial scope in the facilities' extent of environmental management and to demonstrate a meaningful range of environmental performance, involving noncompliance and overcompliance (EPA, 1999, 1997). Also, this sector permits the analysis to exploit similarities and differences across industrial subsectors. Moreover, the chemical sector is responsible for a significant portion of the nation's industrial output and a significant portion of all wastewater discharges by facilities subject to Clean Water Act regulation, as shown in Table 1.1. Nevertheless, the chemical industry is not necessarily representative of all industrial sectors. Indeed, its unique attributes contribute to our interest in studying it. Some firms in the chemical industry, for example, have demonstrated an interest in promoting pollution reduction and prevention through efforts prompted by the Responsible Care program, which is a voluntary management initiative supported by the American Chemical Council.

Third, the chosen focus on two particular wastewater pollutants—TSS and BOD—facilitates our analysis since these pollutants are far and away the most prominent pollutants in our sample, as demonstrated in Chapter 4. Fortunately, this tight scope need not undermine the policy relevance of our study since the pollutants represent two of the five EPA conventional pollutants, which are the focus of EPA efforts. Moreover, nearly all previous studies of wastewater discharges examine exclusively TSS, BOD, or both (for example, Laplante and Rilstone, 1996; Helland, 1998a, 1998b; Earnhart, 2004b, 2004c).

2.3. SURVEY OF ALL PERMITTED FACILITIES
IN THE CHEMICAL MANUFACTURING SECTOR

In order to understand the actions taken by permitted wastewater-discharging facilities to comply with their discharge limits, we designed and implemented a survey of the identified facilities in the chemical manufacturing industry. The survey gathers data on environmental management practices employed by individual facilities, as well as data on the characteristics of these facilities (for example, number of employees). The survey also explores facilities' perspectives on the effectiveness of inspections and enforcement actions at inducing regulated facilities to comply with their NPDES discharge limits. The appendix to Chapter 4 provides a full listing of all the survey questions analyzed in this book.

2.4. IMPLEMENTATION OF SURVEY

The following section describes our implementation of the survey.

2.4.1. Survey Sample

To implement our survey of chemical manufacturing facilities, we first identified the proper population of facilities to survey. The population is based on a full extract drawn from EPA's Permit Compliance System (PCS) database as of September 2001. Our extract contains 2,596 chemical facilities, including both major facilities and minor facilities. For the purposes of implementing the NPDES system, EPA classifies each regulated facility as either "major" or "minor." For this classification, EPA calculates a major rating on the basis of toxic pollution potential, flow type, conventional pollutant load, public health impact, and water quality impact; any discharger with a point total of 80 or more is a "major facility."

To remain in the survey population, facilities needed to meet the following criteria: (1) possessed an NPDES permit, (2) faced restrictions on their wastewater discharges, (3) discharged regulated pollutants into surface-water bodies, (4) were operating as of 2002, and (5) had contact information available from either EPA or alternative sources, for example, phone books. In order to focus on discharges into surface water, we eliminated 984 facilities from the sample because they were either (1) industrial users, that is, users that discharge into pretreatment programs run by publicly owned treatment works; (2) facilities using general permits, for example, gravel pits; (3) facilities discharging only biosolids, that is, sludge; or (4) facilities discharging only storm water. We eliminated industrial users because their discharges are monitored by the publicly owned treatment works into which they discharge. We chose to focus on the direct relationship between state and federal regulators and regulated facilities. We eliminated facilities possessing only general permits because this type of permit involves distinctively different restrictions. By eliminating those facilities discharging only sludge or only storm water, our analysis is better able to examine a sufficiently comparable set of environmental management concerns. In the end, application of all identified criteria identified 1,003 facilities to contact.

2.4.2. Survey Instrument

After identifying the proper population to sample, a research team—consisting of the two authors, Donald Haider-Markel (a political scientist), and Tatsui Ebihara (an environmental engineer)—developed the survey instrument with the assistance of the Policy Research Institute Survey Research Center, which is located within the University of Kansas, and Mark Cohen, the director of the Vanderbilt Center for Environmental Management, working as a hired consultant.[1] Our team pretested the survey instrument with a representative sample of 20 facilities in the Kansas City metro area. Comments during the pretest and the input of Dr. Cohen led to changes in the wording of several questions. However, participants in the pretest expressed no concerns with the length of the survey, and the participants' ease of completion suggested that the survey design was sound.

2.4.3. Survey Methods

The Policy Research Institute Survey Research Center began implementing our survey on March 18, 2002. The survey remained in the field through March 11, 2003.

Survey procedures were as follows. First, initial contact was attempted with the facility. Surveyors asked to speak with either the NPDES contact person listed or whoever was responsible for the facility's regulated water discharge permit. Second, once the correct person was identified and contacted, he or she was informed of our research project and asked to participate in the survey by establishing an appointed day and time to complete the survey. If the potential respondent showed some indication of a willingness to participate, she or he was faxed a short form of the survey that included a number of questions for which the respondent may have needed time to prepare a response, such as: "how many full-time equivalent employees did your facility have in 2001?" Many potential participants were not willing to establish an appointment to complete the survey until after they had reviewed the faxed short form. This request typically meant that several more calls had to be made to the same potential respondent before an appointment could be established.

2.4.4. Response Rate and Assessment of Sample Selection Bias

The Policy Research Institute Survey Research Center contacted 1,003 facilities. Of those facilities contacted, 736 refused to participate in the survey, while 267 facilities completed at least 90 percent of the survey, implying a 27 percent response rate. This rate is comparable to previous large-scale surveys of industrial sectors (for example, Arimura, Hibiki, and Katayama 2008; Delmas and Toffel, 2008).

Nevertheless, given the survey's nonresponse rate of 73 percent, the potential for sample selection bias is a valid concern. Such a bias exists when the set of respondents systematically differs from the original population in meaningful dimensions. As the initial assessment of this concern, we compare the original sample of 1,003 potentially eligible facilities to the 267 facilities that actually completed the survey. Based on this comparison, we find no systematic state or regional bias in survey participation. For example, only the Midwest region is slightly overrepresented in the response group, and only the Northeast region is slightly underrepresented. These differences, however, are small. In addition, across most of the states, the difference between representation in the original sample and representation in the response group averages less than 2 percent. In contrast, our initial assessment reveals some difference in the participation of major facilities versus minor facilities. In the original sample, 69 percent of facilities are minor facilities and 31 percent are major facilities. In the group of survey respondents, major facilities are slightly overrepresented at 39 percent. This difference proves statistically significant.

As a stronger assessment, we test for sample selection bias using the Heckman two-stage sample selection procedure (Heckman, 1979). As the first stage of this procedure, we assess whether any relevant factors appear to affect a facility's decision to complete our survey once the facility is contacted.[2] This assessment reveals a bias in a single dimension: major facilities were more likely to respond to the survey than were minor facilities. Put differently, the Heckman analysis indicates that only the distinction between minor and major facilities proves important for explaining whether or not a contacted facility completed the administered survey. The Heckman analysis demonstrates that neither the preceding history of inspections nor the preceding history of enforcement actions against a particular facility explains whether or not a contacted facility responded to the survey. Moreover, the analysis demonstrates that the decision to respond is not explained by the EPA region in which a particular facility resides. Thus, even if the threat of inspections and enforcement actions varies across EPA regions, this variation does not explain whether or not a contacted facility responds to the survey. (The analysis is not able to control for variation across states in a similar fashion given the large number of individual states relative to the sample size.)

Thus, based on our analysis, it appears that a sample selection bias exists in only a single dimension: the distinction between a "major facility" and a "minor facility." This single distinction proves irrelevant for our final sample of analysis, as explained in the next section. Therefore, in the end, our study does not correct for any potential sample selection bias. This lack of correction is consistent with recent prominently published studies of environmental management practices (Anton, Deltas, and Khanna, 2003; Arimura, Hibiki, and Katayama 2008).

2.5. AVAILABILITY OF DATA ON DISCHARGE LIMITS AND DISCHARGE MEASUREMENTS

As noted earlier, 267 facilities responded to our survey. Of these 267 respondent facilities, 103 facilities are classified as major facilities by EPA; the remaining 164 facilities are classified as minor facilities. EPA's PCS database systematically records wastewater discharges and discharge limits only for major facilities. Not surprisingly, the PCS database contains no information on discharge limits and wastewater discharges for any of the respondent minor facilities. Of the 103 respondent major facilities, the PCS database contains records that potentially provide data on discharge limits and wastewater discharges for 97 major facilities. For all of our analysis, we focus exclusively on these 97 facilities for which we possess both survey data on environmental management practices and EPA data on limits and discharges. (Since EPA records information on discharge limits and wastewater discharges for many facilities, we did not attempt to gather information on these factors using our survey, focusing instead on data elements not recorded by EPA, especially environmental management practices.)

To be clear, this sample includes only major facilities. Our exclusive focus on major facilities clearly limits the generalizability of our results. Nevertheless, our analysis remains policy relevant since EPA focuses its regulatory efforts on EPA-classified "major" facilities. Moreover, major facilities represented 21 percent of the 2,481 chemical facilities in EPA's

NPDES in 2001. Given their size, we suspect that major facilities were responsible for the bulk of wastewater discharges from this sector during the sample period.

As noted earlier, we implemented our survey by contacting facilities between March 2002 and March 2003. Using the survey instrument, we extracted data on environmental behavior (that is, environmental management practices) using three different time formats. For each form of environmental behavior, we chose the most appropriate time format based on the nature of the particular environmental management practice. We measured some forms of environmental behavior during each of the three calendar years between 1999 and 2001, inclusively. For example, we instructed facilities to report within our survey the number of environmental audits on an annual basis between 1999 and 2001. We measured other forms of environmental behavior over the three-year period preceding the completion of the survey. For example, we instructed facilities to report within our survey the amount of money spent on external consultants over the preceding three-year period. We measured the remaining forms of environmental behavior over the twelve-month period preceding the completion of the survey.

Thus, based on these three measurement formats used within our survey instrument, we gathered information on facilities' environmental behavior over the period January 1999 through March 2003. We include the last month in which a facility completed the survey since any surveyed facility may have included the current month within its understanding of the "preceding twelve-month period" or "preceding three-year period."

The time period between January 1999 and March 2003 contains fifty-one months. For our analysis of discharge limits and wastewater discharges, we gather monthly data from the EPA PCS database over the same time period. Thus, our full sample includes 97 facilities, with 51 monthly observations per facility, containing 4,947 facility-month observations. (The appendix to Chapter 5 describes the sample used to examine behavior for each time frame explored.)

This chapter describes our reasons for selecting the chemical industry as the focus of our investigation into regulatory limits, environmental behavior, environmental performance, and government interventions under the Clean Water Act. It also describes the nature and method of implementing a survey we conducted of facilities in that industry during 2002 and 2003 to gain insights into the facilities' environmental management practices and perspectives on the effectiveness of government enforcement actions. The systematic manner in which regulated facilities and regulators record information concerning wastewater discharges, EPA's strong historical interest in the chemical manufacturing sector, and the large component of all industrial discharges for which the chemical industry is responsible contributed to our decision to direct our empirical inquiry at the chemical industry. The analysis of the chemical industry in this book should provide insights into environmental behavior, environmental performance, and the efficacy of government interventions concerning other industries regulated under the Clean Water Act. The chemical manufacturing industry is not necessarily representative of all industrial sectors, however, so policy makers should take into account relevant differences among industries before extrapolating the findings in this book to other industrial sectors.

CHAPTER THREE

Summary of Research Questions and Review of Previous Studies

In this chapter, we summarize our book's research questions and then review previous studies that attempt to answer similar research questions or at least explore the elements relating to our research questions. In this way, we attempt to demonstrate that our book builds on a rich literature of previous studies exploring the connections among effluent limits (for example, wastewater discharge limits), environmental behavior, environmental performance, monitoring, and enforcement.

3.1. SUMMARY OF RESEARCH QUESTIONS

In the first part of this chapter, we summarize our research questions. Some of our research questions are broad in scope. Our first broad research question concerns discharge limits. Simply put, what is the variation in Clean Water Act discharge limits that are imposed on permitted wastewater-discharging facilities embedded within the National Pollutant Discharge Elimination System (NPDES) program? Our second broad research question concerns the actions that wastewater-discharging facilities are taking to comply with these limits; hereafter we refer to these actions as "environmental behavior." Similar to our broad question concerning discharge limits, we pose a simple question: what is the environmental behavior of permitted wastewater-discharging facilities? Our third broad research question concerns the outcomes—in terms of discharges and compliance—of these actions in the presence of the imposed discharge limits; hereafter, we refer to these outcomes as "environmental performance." Highly similar to our broad question concerning environmental behavior, we ask this basic question: what is the environmental performance of permitted wastewater-discharging facilities? Our fourth broad research question concerns the actions taken by regulators to induce compliance with the imposed discharge limits. Our general question is the following: what are regulators doing to induce environmental behavior that effectively controls discharges so that the discharges comply with the imposed limits? Our specific question is the following: what government interventions, namely, inspections and enforcement actions, are regulators taking in order to induce compliance?

In addition to these broad research questions, we pose research questions that explore the relationships connecting discharge limits, environmental behavior, environmental performance, and government interventions. This additional set of research questions is more focused in scope. Of these additional questions, the more important research questions focus on the effects of government interventions on environmental behavior and environmental performance. In particular, we pose these two highly policy-relevant questions: Do the identified government interventions help to induce better environmental behavior? Do the identified government interventions help to induce better environmental performance? Other research questions focus on the effects of discharge limits on environmental behavior and environmental performance. Specifically, we pose these two policy-relevant questions: How do the imposed discharge limits affect environmental behavior? How do the imposed discharge limits affect environmental performance? The final research question focuses on the effect of environmental behavior on environmental performance: does better environmental behavior improve environmental performance?

3.2. REVIEW OF PREVIOUS STUDIES

In the second part of this chapter, we review previous studies that attempt to answer similar research questions or at least explore the elements relating to our research questions.

3.2.1. Previous Studies of Effluent Limits

First, a number of previous studies explore effluent limits. These studies examine elements that relate to the research questions exploring discharge limits.

Some previous studies of effluent limits explore the regulatory process of establishing subindustry effluent limitation guidelines or their equivalent. For example, Magat, Krupnick, and Harrington (1986) examine EPA's efforts to establish subindustry effluent limitation guidelines as part of the implementation of the Clean Water Act. This examination focuses on the two pollutants serving as this book's focus: total suspended solids (TSS) and biological oxygen demand (BOD). Some related studies explore regulatory decisions of pollution control that do not represent quantitative performance-based standards. For example, Cropper et al. (1992) examine the regulatory decision of whether or not to remove certain pesticides from the market.

Other studies explore the determination of plant-specific effluent limits. For example, DeShazo and Lerner (2004) explore the determination of wastewater discharge limits imposed by NPDES permits issued to facilities operating in the pulp, paper, and paperboard industry. Lanoie, Thomas, and Fearnley (1998) explore effluent limits imposed on Canadian pulp and paper mills. In the process of estimating the effect of effluent limits on Canadian pulp and paper mills' abatement investment, wastewater flow, and wastewater discharges, the authors assess the determination of effluent limits by estimating a functional relationship for effluent limit levels. The estimation of this functional relationship is limited to a very small set of explanatory factors: plant-specific production capacity and a set of water basin–specific indicators. Mickwitz (2003) extensively discusses the imposition of permitted discharge limits on pulp and paper mills operating in Finland. The study estimates the regulatory decision of whether to assign a permit limit for a particular substance, in this case, phosphorus. While the study describes decisions to adjust limit levels, it does not assess the decision about a permit limit's level. The study lastly examines the effect of limit levels on BOD discharges and phosphorus discharges. Indirectly, it explores the effect of a limit's presence by comparing the discharges by plants facing no phosphorus limits and discharges by plants facing phosphorus limits. Brännlund and Löfgren (1996) estimate the effect of effluent limits on discharges from Swedish pulp mills. The study only briefly discusses the process of imposing these limits and does not assess the determination of limits as part of the estimation of discharges.

Two related studies by Berman and Bui (2001a, 2001b) examine the effect of air-quality regulation on two broad forms of facility-level environmental behavior: pollution-control investment and operating costs. In particular, these studies examine the effect of air-quality regulation on employment, productivity, air-pollution-control investment, air-pollution-control operating costs, and value added, with a focus on the heavily regulated oil refineries located in the Los Angeles South Coast Air Basin. Both studies measure the increased presence of air-quality regulation based on the count of new regulations announced but not yet imposed, the count of new regulations imposed, and the count of existing regulations that are tightened. None of these regulations are specific to a facility. Instead, they apply to a particular population of facilities depending on the facilities' location, industry, production process (for example, use of a gas-fired steam generator), and legacy (that is, existing source versus new source). Moreover, neither study attempts to measure the actual level of regulatory stringency imposed on an individual facility; instead, the studies focus on changes to the regulatory climate. Furthermore, neither study attempts to explain the introduction of new regulations or the tightening of existing regulations. Also, neither study attempts to examine the effect of air-quality regulation on facility-level environmental performance. (However, both studies demonstrate an apparent relationship between air-quality regulation and emissions by contrasting aggregate emissions between different regions of the United States and the trends over time.) Still, these studies examine the effect of air-quality regulation on a broad form of facility-level environmental behavior: pollution-control investment and operating costs. However, these studies do not attempt to examine the specific dimensions of this pollution control.

3.2.2. *Previous Studies of Environmental Behavior*

Some previous studies explore facilities' environmental behavioral decisions. These studies examine elements that relate to the research questions exploring wastewater-related environmental behavior.

In the broadest set, several previous studies examine the actions taken by polluting facilities to comply with regulatory constraints, along with other forms of environmental behavior. Bluffstone and Sterner (2006) examine firms' decisions to adopt certain components of an environmental management system (EMS), such as the implementation of environmental audits, certification with the protocol developed by the International Organization for Standardisation (ISO), denoted as the ISO 14001 series, which specifies a framework for designing an effective environmental management system, and use of internal water and air pollution monitoring systems. Their study examines industrial firms operating in central and eastern European countries as of 1998. Evans, Liu, and Stafford (2008) examine decisions by Michigan facilities to audit their operations over the years 1998 to 2003. Frondel, Horbach, and Rennings (2007) examine facilities' decisions to install end-of-pipe technologies in contrast to the implementation of cleaner production technologies; for this examination, they examine responses to a 2003 survey implemented by the Organization for Economic Cooperation and Development in several developed countries. Using data from this same survey, with a focus on manufacturing facilities operating in Hungary, Henriques and Sadorsky (2006) examine eight particular environmental management practices, such as decisions to implement an internal audit, implement an external audit, and train facility employees; again these authors also examine the count of environmental management practices implemented. Khanna et al. (2007) examine decisions made by Oregon facilities in 2004 to adopt a certain number of environmental management practices from a set of ten practices. Similarly, Khanna and Anton (2002) and Anton, Deltas, and Khanna (2003) examine decisions made by S&P 500 firms in 1989 and 1990 to adopt a certain number of environmental management practices from a set of thirteen practices. Dasgupta, Hettige, and Wheeler (2000) examine Mexican companies' adoption of an EMS in 1995. Mori and Welch (2008) examine decisions made by Japanese manufacturing facilities in 2001 to adopt an EMS that is ISO 14001 certified, while also distinguishing across "early certifiers," "recent certifiers," and "in-process certifiers." Similarly, Nakamura, Takahashi, and Vertinsky (2000) explore Japanese manufacturing firms' decisions in 1997 whether to adopt an EMS that is ISO 14001 certified. Christmann and Taylor (2001) use a survey of Chinese firms to explore ISO 14001 certification. Delmas and Toffel (2008) explore the adoption of an ISO 14001–certified EMS using data on facilities operating in heavily polluting U.S. sectors in 2003. Collins and Harris (2002) use pollution abatement expenditures to measure the environmental behavior displayed by UK metal manufacturing plants between 1991 and 1994. Similarly, Aden, Ahn, and Rock (1999) examine the annual pollution abatement expenditures of Korean manufacturing plants. Moreover, Wang (2002) examines Chinese factories' extent of environmental behavior as measured by two forms of expenditures: total new investment in wastewater treatment facilities and total operational spending on wastewater abatement. Kerr and Newel (2003) examine the adoption of a specific pollution abatement technique—isomerization— by U.S. refiners between 1971 and 1995. Similarly, Maynard and Shortle (2001) examine decisions to adopt three specific production technologies, arrayed according to their "cleanliness," on the part of U.S. pulp mills between 1987 and 1994. Finally, Snyder, Miller, and Stavins (2003) examine decisions by U.S. chlorine manufacturing facilities to adopt a cleaner production process between 1976 and 2001.

Of these noted studies, some provide analysis that helps to answer this research question: do government interventions help to induce better behavior? Bluffstone and Sterner (2006) examine the influence of wastewater inspections and air pollution inspections on their chosen EMS components. Evans, Liu, and Stafford (2008) examine the influence of inspections and penalties on facilities' audit decisions. Henriques and Sadorsky (2006) and Henriques, Sadorsky, and Kerekes (2005) assess the influence of inspections on a variety of environmental management practices. Khanna and Anton (2002) examine the influence of inspections and civil penalties on the count of environmental management practices implemented by S&P 500 firms.

Dasgupta, Hettige, and Wheeler (2000) assess the influence of regulatory inspections on the extent of Mexican companies' EMSs.

The section above describes the three previous studies that attempt to examine the influence of effluent limits on environmental behavior (Lanoie, Thomas, and Fearnley, 1998; Berman and Bui, 2001a, 2001b), which represents one of our research questions.

3.2.3. Previous Studies of Environmental Performance

Some previous studies explore the environmental performance of regulated entities. These studies examine elements that relate to several of our research questions.

In the broadest set, several previous studies examine the pollution-related outcomes of polluting facilities' actions taken in the presence of regulatory limits; such outcomes include wastewater discharges, air-pollutant emissions, toxic and hazardous waste generation, and oil spills, along with the status or degree of compliance with regulatory limits. Anton, Deltas, and Khanna (2003) examine the environmental performance displayed by S&P 500 firms in 1989 and 1990 based on four per-output unit measures: total toxic release inventory releases, onsite discharges of toxic emissions, offsite discharges of toxic emissions, and hazardous discharges. Dasgupta, Hettige, and Wheeler (2000) examine Mexican companies' compliance with environmental regulations in 1995. Barla (2007) examines TSS and BOD wastewater discharges emitted by Quebec pulp and paper plants between 1997 and 2003. Dasgupta et al. (2001) examine wastewater discharges—TSS and chemical oxygen demand—emitted by Chinese manufacturing companies between 1993 and 1997. Gangadharan (2006) examines compliance with environmental regulations on the part of Mexican plants in 1995. Wang and Wheeler (2005) examine chemical oxygen demand wastewater discharges and air-pollutant emissions from Chinese facilities in 1993. Similarly, Wang et al. (2003) examine the level of wastewater discharges (relative to imposed legal limits) from Chinese industrial plants during the years 1993 to 1997. Arimura, Hibiki, and Katayama (2008) examine Japanese facilities' environmental performance as measured by natural resource use, solid waste generation, and wastewater discharge. Evans, Liu, and Stafford (2008) examine Michigan facilities' compliance with air emission limits and hazardous waste regulations. Gray and Deily (1996) examine the air-related compliance decisions of U.S. steel factories. Earnhart and Lizal (2006) examine air pollutant emissions in the Czech Republic during the country's transitional years of 1993 to 1998. Magat and Viscusi (1990) study discharges of BOD from U.S. pulp and paper mills between 1982 and 1985. Laplante and Rilstone (1996) study wastewater discharges by the pulp and paper industry in

Quebec between 1985 and 1990. Foulon, Lanoie, and Laplante (2002) also examine the discharges of Canadian pulp and paper mills. Similarly, Shimshack and Ward (2005) examine wastewater discharges from U.S. pulp and paper mills. Moreover, Nadeau (1997) examines the duration of noncompliance with air-pollution control regulations on the part of U.S. pulp and paper mills. In addition, Helland (1998a, 1998b) examines wastewater discharges from U.S. pulp and paper mills. Gray and Shadbegian (2005) examine the likelihood of U.S. pulp and paper plants complying with air-pollution regulations. Earnhart (2004a, 2004b, 2004c) examines wastewater discharges from municipal wastewater treatment plants operating in Kansas between 1990 and 1998. Stafford (2002) examines the compliance status of hazardous waste facilities. The final set of previous studies examines accidental discharges (for example, oil spills). Epple and Visscher (1984), Anderson and Talley (1995), Cohen (1987), and Viladrich-Grau and Groves (1997) examine spills in the oil transport sector. Eckert (2004) examines oil storage violations in Canada.

Of these noted studies, several provide analysis that helps to answer this research question: do government interventions help to induce better performance? Stafford (2002) examines the influence of inspections on the compliance status of hazardous waste facilities; her findings show that increased inspections against a specific facility do not influence compliance status and that increased inspections at other facilities in the relevant state, which represents a measure of monitoring threat, significantly lowers compliance status. Helland (1998a) examines the influence of inspections on compliance with BOD discharge limits imposed on U.S. pulp and paper mills, while Helland (1998b) examines the influence of inspections on absolute BOD discharges; both of his findings reveal that inspections against specific facilities do not influence environmental performance regardless of its form. Gray and Deily (1996) examine the influence of inspections and "other enforcement actions" on steel factories' air-related compliance decisions. Their findings reveal that lagged government interventions—as measured by the sum of inspections and enforcement actions—against a specific facility increase the likelihood of a steel mill's compliance; the same conclusion holds for only inspections. However, the authors find that an increased threat of government intervention from a mix of inspections and enforcement actions does not influence the likelihood of compliance; the same conclusion holds for only inspections. Gray and Shadbegian (2005) examine the influence of threatened inspections and "other enforcement actions" on the likelihood of pulp and paper plants' complying with air pollution regulations; their findings demonstrate that an increase in the threat of "other enforcement actions" improves the likelihood of compliance. Laplante and Rilstone (1996) examine the influence of inspections on BOD and TSS absolute discharges by Quebec pulp and paper plants and the ratios of discharges to imposed limits. Their findings show that lagged inspections at the specific facility reduce BOD discharges and the ratio of BOD discharges to the imposed limit. Moreover, the authors demonstrate that an increased threat of inspections decreases TSS absolute and relative discharges and BOD absolute discharges. Magat and Viscusi (1990) examine the influence of inspections on BOD absolute discharges from U.S. pulp and paper mills and their compliance status with imposed discharge limits; their findings reveal that lagged inspections against the specific facility lower absolute discharges and decrease the probability of noncompliance. Earnhart (2004c) analyzes BOD-related wastewater discharges (relative to discharge limits) from publicly owned treatment works. His findings reveal the following: (1) federal and state inspections against specific facilities are ineffective at lowering discharges; (2) an increased threat of federal inspections significantly lowers relative discharges, but an increased threat of state inspections does not significantly affect relative discharges; (3) both the threat of state enforcement and the threat of federal enforcement negatively influence relative discharges; (4) lagged enforcement against a specific facility significantly lowers relative discharges. Shimshack and Ward (2005) examine the influence of inspections and enforcement actions on compliance with wastewater discharge limits. Their findings show that the threat of enforcement as demonstrated by actions taken against other similar facilities improves a particular facility's compliance status; however, their findings also demonstrate that lagged enforcement against a specific facility does not influence compliance status. Nadeau (1997) examines the influence of inspections and enforcement on compliance with air regulations. His findings show that an increased threat of enforcement reduces the duration of noncompliance, yet an increased threat of inspection does not generally influence noncompliance duration. Foulon, Lanoie, and Laplante (2002) examine the influence of enforcement on wastewater discharges by considering the prosecutions performed against and fines imposed on a facility in a given year; the authors find no evidence of enforcement affecting discharges. Epple and Visscher (1984), Cohen (1987), Anderson and Talley (1995), and Viladrich-Grau and Groves (1997) examine the effect of government inspections on oil spills. These studies reveal that Coast Guard inspections lower the volume of oil released in oil spills. Viladrich-Grau and Groves (1997) additionally demonstrate that such inspections reduce the frequency of oil spills. Viladrich-Grau and Groves (1997) also examine the influence of an increased threat of enforcement on oil transport spills; their findings show that the threat of enforcement influences neither the volume of oil spilled nor the frequency of oil spills. Eckert (2004) examines the effect of enforcement actions on Canadian oil storage violations; her findings reveal that lagged enforcement against a specific facility does not affect the presence of violations.[1]

In addition, in the broadest set of studies, some studies provide analysis that helps to answer this research question: how does environmental behavior influence environmental performance? Anton, Deltas, and Khanna (2003) examine the influence of the comprehensiveness of a firm's EMS, as measured by the count of adopted environmental management practices, on four measures of a firm's toxic emissions per unit of output; the study demonstrates that the extent of EMS adoption has a significant negative impact on the intensity of toxic emissions. Dasgupta, Hettige, and Wheeler (2000) examine the influence of a Mexican company's EMS on the company's environmental compliance status; the study finds that the adoption of an ISO 14001–certified EMS leads to a significant improvement in compliance status. Barla (2007) examines the influence of ISO 14001 certification on wastewater discharges from Quebec pulp and paper plants. The study tests whether adopting an ISO 14011–certified EMS affects environmental performance; results reveal that adoption decreases BOD discharges but not TSS discharges. Arimura, Hibiki, and Katayama (2008) explore the effect of ISO 14001 implementation on Japanese facilities' environmental performance (as measured by natural resource use, solid waste generation, and wastewater discharge); results demonstrate that ISO 14001 implementation reduces all three impacts. Evans, Liu, and Stafford (2008) examine the influence of facilities' environmental self-audits on compliance with environmental regulations; results suggest no long-run impact of auditing on compliance with air emission limits but a positive effect on future compliance with hazardous waste regulations.[2]

Lastly, Section 3.2.1 (see earlier in this chapter) describes the few previous studies that attempt to examine the influence of effluent limits on environmental performance (Lanoie, Thomas, and Fearnley, 1998; Mickwitz, 2003; Brännlund and Löfgren, 1996), which represents one of our research questions.

3.3. SUMMARY

This chapter identifies the research questions this book addresses, as well as the existing empirical literature concerning effluent limits, environmental behavior, and environmental performance. The chapter also describes the scope and limits of previous empirical work on environmental enforcement. This brief discussion places the empirical findings in this book in a larger context, allowing readers to determine whether these findings are consistent with earlier empirical work and how this book addresses gaps in the existing environmental enforcement literature.

Discharge Limits Imposed on Discharging Facilities

4.1. AUTHORITY TO IMPOSE DISCHARGE LIMITS UNDER THE CWA

The Clean Water Act, like most federal pollution-control statutes, is an intricate and complex law. Its overarching goals and the principal legal mechanism for achieving them, however, are simple, if not elegant. The statute establishes a goal of eliminating discharges of pollutants into waters of the United States and an interim goal of achieving fishable, swimmable waters, wherever attainable.[1] The principal mechanism for achieving these goals is the statute's stark enunciation that, with certain exceptions, "the discharge of any pollutant by any person shall be unlawful."[2] Despite the apparent breadth of this prohibition, the Clean Water Act's bite is considerably weaker than its bark. The statute created two permit programs and exempts pollutants discharged in compliance with a valid permit under either program from the statutory prohibition on discharges. This book focuses on the first permit program, the National Pollutant Discharge Elimination System (NPDES) permit program, which covers discharges of industrial and municipal effluents into surface-water bodies.[3] The other program, the Section 404 program, authorizes discharges of dredged or fill material into navigable waters.[4] The Section 404 program, which applies most frequently to deposits of materials into wetlands, is not the subject of the analysis in this book.

The Clean Water Act authorizes the administrator of the federal Environmental Protection Agency (EPA) to issue NPDES permits for the discharge of any pollutant, notwithstanding the statutory prohibition on discharges.[5] A permit must require that its holder meet a set of requirements listed in the statute. Among these is the requirement to comply with effluent limitations that the Clean Water Act authorizes the EPA to adopt. An effluent limitation is a restriction on quantities, rates, and concentrations of chemical, physical, biological, or other constituents discharged by point sources into navigable waters.[6] Although the Clean Water Act uses the term "effluent limitations" broadly to include restrictions on discharge contained in both regulations and permits, we distinguish throughout this book between "effluent limitations" and "discharge limits." The former are contained in EPA regulations described in Section 4.1.3 later in this chapter. The latter are enforceable restrictions included in individual discharge permits issued either by EPA or by states authorized to administer the NPDES permit program. The discharge limits contained in each NPDES permit differ from point source to point source, depending on a variety of factors discussed in this chapter. NPDES permits typically also impose record-keeping and reporting requirements on their holders and subject holders to facility inspections by federal or state regulators.[7] Unless it is impracticable, permit limits must be expressed as both average monthly limits and maximum daily limits (EPA, 1996). Throughout this book, we examine data on only the average monthly limits because the available data on maximum daily limits is incomplete and may be inaccurate.

Permit holders found to have violated applicable discharge limits or other permit conditions are subject to civil or criminal sanctions. Our focus in this book is on civil enforcement by EPA and the states (although the Clean Water Act also allows private individuals and nongovernmental organizations to enforce permits in certain circumstances in "citizen suits"). Under a provision known as the "permit shield" provision, any point source that complies with a valid NPDES permit is protected from enforcement actions against it by EPA, the state that issued the permit, and individuals or organizations bringing citizen-suit enforcement actions.[8]

4.1.1. EPA Regional Authorities

The responsibility for implementing and enforcing the Clean Water Act is multilateral, with EPA and the states holding the biggest sticks. The national headquarters of the EPA in Washington, D.C., is primarily responsible for the adoption of regulations containing effluent limitations. The responsibility for issuing NPDES permits that impose those discharge limits on particular dischargers falls to the agency's regional offices. EPA has ten regional offices located throughout the country. The regional offices are also responsible for enforcement of statutory and regulatory provisions and for the provision of compliance assistance to regulated entities. Although a vigorous debate has developed over the relative merits of deterrence-based enforcement and compliance assistance as government-wielded tools for increasing compliance with environmental laws such as the Clean Water Act, our interest here lies in the pursuit of deterrence-based enforcement.

4.1.2. State Primacy

Although the Clean Water Act delegates to EPA the responsibility to administer the NPDES permit program, it also reflects the commitment to "cooperative federalism" that animated Congress when it adopted many of the nation's leading environmental laws in the early 1970s. The statute authorizes the states to seek permission to take over the NPDES program as it applies to point sources that are discharging pollutants within their jurisdiction. By 2003, EPA had provided at least partial approval of state NPDES permit programs to at least forty states and territories. In those jurisdictions lacking an approved program, EPA continues to administer the NPDES program.

The governor of any state desiring to administer the NPDES permit program for discharges within the state's jurisdiction may submit to EPA's administrator a description of the program it proposes to establish and administer under state law. The state's attorney general or the independent legal counsel for its water-pollution control agency also must attest that state laws provide adequate authority to implement the program.[9] The Clean Water Act requires that EPA approve a state program unless it determines that adequate authority does not exist to perform certain functions. These include the usual suspects, such as authority to apply and ensure compliance with the EPA's effluent limitations; to inspect, monitor, and require reports from dischargers; and to halt permit violations, including through the imposition of civil and criminal penalties.[10]

Once EPA approves a state NPDES permit program, EPA suspends the issuance of federal permits for activities subject to the approved state program.[11] In such cases, the state has "primacy" over administration of the NPDES permit program. Even in states with primacy, however, EPA retains veto power over individual state permits.[12] EPA retains concurrent authority to enforce the statute and EPA regulations and state permits issued under it.[13] Permit holders are therefore answerable to the federal government for exceeding discharge limits and committing other regulatory violations, even if they receive their permits from a state environmental agency.

4.1.3. *Effluent Limitation Guidelines derived by EPA*

The Clean Water Act vests in EPA the authority to issue effluent limitation guidelines as part of the process of issuing enforceable discharge limits. Concerned about the time it would take to issue individualized discharge limits to each point source discharging into the nation's surface waters, EPA decided shortly after the statute's enactment to adopt effluent limitations in the form of regulations that apply to *classes* or *categories* of point sources in particular industries. The U.S. Supreme Court declared this practice valid, concluding that the act authorizes EPA to adopt enforceable effluent limitations in the form of regulations.[14] As a result, EPA typically issues regulations that include the effluent limitation guidelines for a particular category of point sources. The permit-issuing agency (whether that is a state with primacy or EPA in a state that lacks permission to administer the NPDES permit program) generally determines the appropriate discharge limit for a particular point source by incorporating into an individual discharge permit the effluent limitations assigned in the regulations to the relevant industrial category. A point source applying for a discharge permit may request a different facility-specific discharge limit by applying for a variance or modification, which, if granted, tailors the discharge limit to the particular circumstances of the applicant. The availability and significance of these individualized adjustments are discussed in Section 4.2.3 later in this chapter.

When EPA sets effluent limitations by regulation for a particular industrial category, it must consider the factors specified in the Clean Water Act. The statute provides for two phases of discharge control of increasing stringency, with each phase advancing toward the statutory goals of achieving fishable-swimmable waters and, ultimately (at least in theory), the elimination of all discharges. These requirements are referred to as the Clean Water Act's technology-based pollution controls because EPA sets the effluent limitation guidelines in the regulations based on what it deems to be technologically and economically feasible for the particular category of dischargers covered by the regulations. These technology-based effluent limitation guidelines differ from water-quality standards, which prescribe minimally adequate levels of receiving water quality. As described later in the chapter, effluent limitation guidelines may or may not differ from the discharge limits imposed in an NPDES permit for an individual discharger in the industrial category covered by the regulations.

Although this point is often misunderstood, or glossed over by those seeking to disparage the Clean Water Act as an example of onerous, overly prescriptive, ill-advised, and meddlesome command-and-control regulation, it is important to recognize that the technology-based effluent limitations are performance standards, not design (or specification) standards (Glicksman and Earnhart, 2007b). The distinction is critical. If an environmental agency relies on a *design standard* to achieve some statutory environmental protection goal, it defines the precise method by which regulated entities must achieve the goal, such as by installing and operating a particular kind of pollution-control technology or work practice identified and mandated by the agency. That is not the way the Clean Water Act works. EPA derives the regulatory limitations by determining the level of pollution control that is feasible for a particular category of point sources to achieve. Dischargers are then free to select the means they will use to comply with the discharge limits derived from those regulatory limitations and to which they are subject. Some sources assuredly will use the model technology or work practice identified by the agency as the one that makes compliance possible. Sources are free, however, to devise alternative means of complying with their discharge limits. These kinds of *performance standards* are often regarded as preferable to design standards because they provide dischargers with an incentive to develop means of complying with applicable discharge limits that are less costly or more effective than the ones EPA relied on in setting the regulatory limitations (Shapiro and Glicksman, 2000).

The Clean Water Act did not seek to transform the nation's surface waters, which often were highly polluted when the statute was adopted, into pristine waters in one fell swoop. Rather, the statute envisions step-by-step movement toward the ultimate, no-discharge goal. All industrial point sources had to comply initially with effluent limitations reflecting the application of the best practicable control technology currently available, no matter what kind of pollutant they discharged.[15] For most industries, the second phase of controls kicked in during the 1980s and continues to apply today. This second phase of controls is more complex (and requires familiarity with more acronyms) than the first. The nature of the limitations applicable to a particular category of industrial point sources at this second phase turns on the kind of pollutant being discharged. Point sources discharging conventional pollutants must comply with limitations that reflect the best conventional pollutant control technology.[16] Conventional pollutants include pollutants classified as biological oxygen demanding, suspended solids (or total suspended solids), fecal coliform, and pH.[17] Point sources that discharge either toxic pollutants or nonconventional, nontoxic pollutants must comply with effluent limitations that reflect the application of the best available technology economically achievable.[18]

All of the regulatory effluent limitation guidelines are supposed to reflect a commitment to forcing point sources to achieve technologically achievable levels of control. The levels of control are expected to become more stringent over time so that industry moves closer to the aspirational no-discharge goal. Generally, the least stringent regulatory limitations apply to discharges of conventional pollutants and the most stringent controls apply to discharges of toxic pollutants, which are regarded as the most dangerous for both public health and the environment.

As if this phased system of discharge reductions is not confusing enough, an entirely different set of requirements applies to newly constructed industrial point sources. Although these requirements are also premised on an assessment of how much discharge control is technologically possible within a particular category of point sources, in general, the requirements that apply to new sources are supposed to be more stringent than those that apply to existing point sources in the same category. The theory for making controls for new sources more stringent is that industry is typically able to build pollution-control technology into a newly constructed source more efficiently than if it retrofits sources that began operating before the adoption of the relevant pollution controls. In practice, however, EPA often equates its regulatory standards for new sources with the second phase of the regulatory effluent limitation guidelines applicable to the same category or subcategory of industrial point sources. It is that second phase of effluent limitations (and resulting individualized discharge limits) that is most relevant to discharges by chemical manufacturing point sources during the period covered by our study.

EPA has issued regulatory effluent limitation guidelines for the chemical manufacturing industries. One set of guidelines applies to organic chemicals, plastics, and synthetic fibers, with additional subcategories.[19] The guidelines include separate provisions for direct-discharge point sources that use end-of-pipe biological treatment, direct-discharge point sources that do not use end-of-pipe biological treatment, and indirect-discharge point sources (that is, those that send their waste to public sewage treatment facilities instead of discharging them directly into surface-water bodies; our study does not cover these indirect dischargers). The point sources covered by these regulations are included within Standard Industrial Classification (SIC) Codes 2821, 2823, 2824, 2865, and 2869.[20] As is often the case, EPA's regulations are extensive. The guidelines for the inorganic chemical manufacturing point source category apply to at least forty-seven different production subcategories.[21] This multiplicity of guidelines for a single industry precludes us from identifying the guidelines that apply to the individual facilities we examined because we lack the information necessary to do so.

EPA's regulatory effluent limitation guidelines do not always cover every pollutant present in a discharge from a point source in the relevant industrial category or subcategory. Often, EPA establishes guidelines for only a subset of the pollutants discharged by a particular category of point sources. These pollutants are referred to as *indicator pollutants* because EPA has determined that the discharge control methods that point sources rely on to control these pollutants often are capable of adequately controlling other pollutants discharged by those point sources (EPA, 1996). EPA takes the position that it may be appropriate for a permit to contain no discharge limits for a particular pollutant if a discharge level is so low that it will not interfere with designated uses of the receiving water. One government study finds that about 70 percent of roughly 1,400 permits issued to municipal waste treatment plants included neither discharge limits nor monitoring requirements, a somewhat shocking result that seems completely at odds with the statute's enunciation of a no-discharge goal and its general prohibition on all unpermitted discharges (GAO, 1996).

4.2. THE PERMIT-WRITING PROCESS IN WHICH LIMITS ARE IMPOSED

Although EPA regulations specify the effluent limitations for categories of industrial point sources, the mechanism for imposing those limitations on individual point sources is the NPDES permit process. As indicated in Section 4.1.3, EPA's regulations form the starting point for identification of the appropriate discharge limits for an individual point source. The limitations reflected in the regulations may or may not wind up being the discharge limits imposed on a point source in its permit, however. This section first describes the process by which point sources apply for NPDES permits and EPA and the states' rules on permit applications. It then explains why point sources from the same industrial category or subcategory may be subject to remarkably different discharge limits in their NPDES permits, notwithstanding the fact that they are covered by the same regulatory effluent limitation guidelines.

4.2.1. NPDES Permit Application Process

The permit process begins when the owner or operator of a point source submits an application for a permit to the director of the appropriate regional office of EPA or to the appropriate state environmental agency in states with primacy. Permits typically contain applicable conditions, compliance schedules,[22] monitoring requirements, and discharge limits.[23] EPA regulations require that every NPDES permit include certain conditions with which permit holders must comply. Permit holders must allow EPA or state permitting agencies to enter the premises where a regulated facility is located or conducted or where required records are kept, to have access to and copy at reasonable times required records, to inspect at reasonable times any facilities or equipment regulated under a permit, and to sample or monitor substances or parameters for the purposes of ensuring permit compliance.[24] Permit-issuing agencies may establish additional conditions on a case-by-case basis to ensure compliance with applicable limits.[25]

Noncompliance with permit conditions is grounds for enforcement action (civil or criminal) and for termination of the permit, modification of the permit, or denial of a permit renewal application.[26] Permits are effective for a maximum fixed term of five years[27] but may be modified, revoked, or terminated for cause.[28]

4.2.2. Technology-Based Versus Water Quality–Related Effluent Limitations: Why Discharge Limits Differ Within an Industrial Category

The discharge limits that are included in NPDES permits are based in part, but not exclusively, on EPA's technology-based, regulatory effluent limitation guidelines. With limited exceptions, every permit must require compliance, at a minimum, with the regulatory effluent limitations specified in EPA's effluent limitation guidelines for existing point sources or standards of performance for new sources.[29] However, discharge limits are affected not only by the regulatory guidelines and standards, but also by the water-quality standards that the Clean Water Act requires the states to adopt and achieve. In particular, point sources must comply either with the regulatory effluent limitation guidelines adopted by EPA or with discharge limits imposed under state law, including limits needed to meet state water-quality standards, whichever are more stringent.[30]

Under Section 303 of the Clean Water Act, every state must adopt and submit for EPA approval a series of state water-quality standards for various pollutants.[31] The statute requires that the states review and, if appropriate, revise those standards at least once every three years.[32] The standards must be sufficient "to protect the public health or welfare, enhance the quality of water and serve the purposes" of the Clean Water Act.[33]

Each state water-quality standard is composed of several elements.[34] Any of these elements may affect the stringency of individual discharge permits. First, the standard must identify a designated use of the water body covered by the standard (such as drinking, recreation, irrigation use, or waste discharge). Use classifications must take into account the water-quality standards of downstream users and ensure that the state's standards provide for the attainment and maintenance of the water-quality standards of downstream waters.[35] The "higher" the use designation, the more stringently point sources discharging into the relevant water body must be regulated to ensure compliance with the state's water-quality standard. For example, a body of water designated for use as a drinking-water source requires cleaner water than one designated for use as irrigation water. The cleaner the receiving water is required to be, the lower the level of aggregate discharges the water body will be capable of assimilating without violating the state water-quality standard. Although states have considerable discretion in identifying designated uses, EPA regulations impose some boundaries, for example, by restricting the circumstances in which states may downgrade uses that are already being achieved.[36]

Second, each water-quality standard must include water-quality criteria for the designated use.[37] These "criteria" are not regulatory polices or prescriptions. Rather, they are documents that reflect scientific information on the effects of discharges of the relevant pollutant on health and welfare and biological community diversity, productivity, and stability.[38] These criteria are designed to ensure that the quality of the water covered by the water-quality standard is suitable for the designated use.

Third, some state water-quality standards have another component called a total maximum daily load (TMDL). The Clean Water Act requires states to identify bodies of water that do not meet applicable water-quality standards, despite full compliance by point sources discharging into those waters with EPA's regulatory effluent limitations and new source standards of performance. For each of these "impaired waters," the state must compute a TMDL, which represents the maximum assimilative capacity of the receiving body of water for the pollutant for which the water is impaired.[39] TMDLs must be set "at a level necessary to implement the applicable water quality standards with seasonal variations and a margin of safety which takes into account any lack of knowledge concerning the relationship between effluent limitations and water quality."[40] The idea is to determine how much of the pollutant may be discharged into the water during a certain period of time without making the water unusable for the designated use specified in the water-quality standard or without exceeding the water-quality criteria contained in the standard. States must ensure that aggregate discharges into the affected body of water, by point and nonpoint sources alike, do not exceed the TMDL. Although states have discretion to determine how to allocate the aggregate discharges identified by a TMDL among dischargers, they must require, at a minimum, that point source discharges comply with applicable technology-based regulatory limitations adopted by EPA.

It is possible that even if all point sources that discharge into a particular water body comply with their individual discharge limits, the quality of the receiving water in that body may still not be adequate to satisfy state water-quality standards. If it is not, permitting agencies must impose on at least some of those sources discharge limits that are more stringent than the effluent limitations in EPA's regulations. These more stringent limits account for some of the variations in the levels of control applicable in NPDES permits to different point sources in the same industrial category. At least one study of Clean Water Act permits documents these variations. The GAO (1996) finds a numeric limit for one facility discharging mercury that is about 775 times greater than the discharge limits contained in a permit for another facility in the same industry with similar capacity. In short, differences in the quality of the water into which discharges flow account for some of the differences in discharge limits among point sources in the same industry. Generally, the higher the level of the background concentration of a pollutant already present in receiving waters, the more stringent the discharge limit must be to avoid violating state water-quality standards for that pollutant (GAO, 1996).

Other factors may account for differences among the discharge limits imposed in NPDES permits on point sources in the same industrial category. Subject to EPA approval,[41] states may include in their water-quality standards policies affecting their implementation, such as mixing zones, low flows, and the potential for dilution (measured by the ratio of the low flow of the receiving waters to the flow of the discharge; GAO, 1996). These policies can affect the stringency of a permit's discharge level. Mixing zones, for example, "are areas where an effluent undergoes initial dilution and are extended to cover secondary mixing in the ambient water body" (EPA, 1996, p. 107). The states have discretion in choosing whether or not to use a mixing zone. Generally, discharge limits are less stringent if a facility is located in a state that allows mixing zones than they would be if the state required facilities to meet numeric limits at the end of the discharge pipe (GAO, 1996). State permitting agencies also use different methods to calculate the operating capacity of point sources. These calculations affect the stringency of discharge limits because EPA's regulatory limitations or new source standards may be expressed in terms of allowable pollutant discharge per unit of production (DeShazo and Lerner, 2004).

EPA expresses most regulatory effluent limitation guidelines and new source standards of performance in terms of either allowable pollutant discharge per unit of production (or some other measure of production) or wastewater flow rates. Ideally, production or flow would be constant over time, so that the limitations could be based on the uniform rate of production or flow. In practice, however, production and flow rates are not constant due to factors such as market demand, breakdowns, maintenance, product changes, downtimes, and facility modifications. Permit writers therefore must adapt the guidelines when they apply them to facilities with varying production or flow rates (EPA, 1996).

In addition, the Clean Water Act preserves states' authority to impose or enforce any discharge limit that is at least as stringent as the ones mandated by EPA regulations.[42] As a result, any state that seeks to achieve clean surface-water quality to a greater degree or at a faster pace than the federal government mandates is free to issue NPDES permits that impose discharge limits on particular point sources that are more stringent than EPA's regulatory effluent limitation guidelines or new source performance standards. One point source therefore may discharge a pollutant into a water body governed by a considerably more stringent water-quality standard than the standard that applies to another body of water into which a point source in the same industrial category discharges the same pollutant. That difference may result in the first point source being subject to more stringent discharge limits than the second. EPA regulations require that each NPDES permit incorporate any more stringent discharge limit established under state law not only because such a limit is required to attain or maintain water-quality standards but also because the state decides as a matter of policy to adopt more stringent controls than the ones in EPA's technology-based regulations.[43]

4.2.3. *Variances and Modifications*

One final reason why discharge limits may differ among point sources in the same industrial category is the issuance of variances or modifications by permitting agencies. The Clean Water Act and EPA's implementing regulations allow an individual point source to request that the permit-issuing agency adopt a discharge limit that is more lenient than the technology-based limitations contained in EPA's regulatory guidelines. Conversely, nongovernmental organizations may request the imposition of discharge limits that are more stringent than the regulatory limitations. The time to request such an individualized limit is when the owner or operator of a point source submits its permit application.[44] If the permit-issuing agency agrees that deviation from the regulatory limitation is justified, the agency will include in the permit a variance or modification. Because our data do not identify whether point sources have received any of the available variances or modifications, we do not discuss them further. In any event, it appears that issuance of such individualized variations is relatively rare.

4.2.4. *Discharge Limits Based on Best Professional Judgment*

EPA has not issued regulatory effluent limitation guidelines or new source standards of performance for all industrial point source categories. If EPA has not issued guidelines for a particular class or category of point sources (whether due to lack of time, resources, interest, or information), permitting agencies must establish the discharge limits that apply to a particular point source within that category on a case-by-case basis (EPA, 1996).[45] A permitting agency sets these facility-specific limits on the basis of its best professional judgment.[46] The authority to craft discharge limits in the absence of applicable regulatory effluent limitation guidelines provides a significant degree of flexibility to NPDES permit writers in determining the appropriate discharge limits, taking into account any unique factors relating to the permit applicant (EPA, 1996). Permitting agencies must set best-professional-judgment discharge limits, however, on the basis of the same kinds of factors that EPA must consider in adopting effluent limitation guidelines for the particular discharges in question, for example, best practicable control technology or best conventional pollutant control technology (EPA, 1996). These factors are generally not within the control of the point sources seeking the permit. Rather, they may include the cost of using a particular control technology or the extent to which a particular discharge limit may create adverse nonwater-quality environmental impacts (such as air pollution).

4.3. DISCHARGES AND DISCHARGE LIMITS

This section describes the kinds of discharges covered by our study and the manner in which EPA and the states measure discharges and determine whether these discharges comply with discharge limits.

4.3.1. *Direct Discharges into Surface-Water Bodies*

Our study focuses on point sources that discharge pollutants directly into surface-water bodies. It does not address industrial users, which discharge into publicly owned treatment works (also known as municipal sewage treatment plants) that are subject to pretreatment programs run by those treatment works. EPA regulations require that each NPDES permit establish discharge limits for each outfall or discharge point within a facility that is required to possess an NPDES permit.[47]

4.3.2. Methods for Measuring Wastewater Discharges and Formulating Discharge Limits

Most regulatory guidelines or new source standards of performance are measured in terms of quantities (for example, pounds per day). However, some are expressed in terms of concentrations, which reflect allowable pollutant discharge per unit of wastewater (for example, milligrams of pollutant per liter of flow). Consistent with this duality, nearly all discharge limits are expressed in terms of quantities or concentrations. However, in a few cases, discharge limits are expressed in terms of visual observations (for example, no visible sheen or floating solids), limitations on indicator parameters, pH ranges, or temperature limits (Gallagher and Miller, 1996).

4.3.2.1. Discharge Limits Based on Quantities or Concentrations. Under EPA's NPDES permitting regulations, any discharge limits that are connected to production (or other measures of operation, such as flow) must be based not on the designed production capacity, but instead on a reasonable measure of actual production at the facility. For new sources, production must be estimated using projections.[48] Because production tends to vary, permit writers determine a single estimate of the long-term average production level that is expected to exist during the term of the permit. The permit writer multiplies this single production value by the effluent limitation guidelines' rate of allowable pollutant discharge per unit of production to calculate quantity-based discharge limits in the permit (EPA, 1996).

If production rates are expected to change significantly over the course of the permit, the permitting agency can establish alternate or tiered limits that become effective when production exceeds a threshold value. Generally, permit writers should consider tiered limits when changes in production are expected to exceed 20 percent (EPA, 1996).

Permitting agencies have the discretion to express discharge limits in more than one dimension (for example, quantity). However, EPA regulations state that all pollutants limited in permits must have discharge limits expressed in terms of mass (such as pounds, kilograms, or grams), except when applicable limitations are expressed in terms of other units of measurement or limitations in terms of mass are infeasible because the mass of the pollutant cannot be related to a measure of operation and permit conditions ensure that dilution will not be used as a substitute for

treatment.[49] Our sample includes many instances in which this requirement is not met. Instead, several facility permits include only concentration limits. The presence of only concentration limits is problematic
because EPA has stated that "expressing limitations in terms of concentrations as well as mass encourages the proper operation of a treatment
facility at all times" (EPA, 1996, pp. 66–67). Concentration limits avoid
the problem that results when the discharged waste contains higher pollutant concentrations than does the receiving water. Still, the presence of
both limit types—quantity and concentration—is important because quantity limits promote water conservation by removing a discharger's ability
to dilute effluent with clean water as a means of compliance. In other words,
the use of concentration limits may not be appropriate if they would dis-
courage the use of innovative treatment techniques, such as water conservation. As a result, the applicability of concentration-based limits is a
case-by-case determination that requires the exercise of the permit writer's professional judgment (EPA, 1996).

When included in facility permits, concentration limits are supposed to
be based on an evaluation of historical monitoring data and the permit
writer's engineering judgment. EPA directs permit writers to use long-
term average flow to calculate concentration limits because this value best
reflects the range of concentrations that can be expected in a well-operated
plant (EPA, 1996).

4.3.2.2. State Water Quality–Related Discharge Limits.   Before the permitting agency may calculate a water quality–based discharge limit that
ensures that the discharge does not result in a violation of the applicable
water-quality standard, the agency must first determine the point source's
waste-load allocation. EPA defines a waste-load allocation as the fraction
of a TMDL for the water body that is assigned to a particular point source
(EPA, 1996). Permitting agencies use water-quality models to establish
waste-load allocations. These models establish a quantitative relationship between a waste's load and its impact on the quality of the water body
into which it is discharged (EPA, 1996). The choice of the model depends
on whether there is rapid and complete mixing of the effluent with the
receiving water. If so, EPA recommends a complete mix assessment involving fate and transport models. If not, EPA recommends a mixing-zone
assessment.

The goal of the permitting agency is to impose discharge limits, whether
based on EPA's regulations or on the need to comply with state water-
quality standards, that are "enforceable, adequately account for effluent variability, consider available receiving water dilution, protect against
acute and chronic impacts, account for compliance monitoring sampling

frequency, and assure attainment of the WLA [waste-load allocation] and water quality standards" (EPA, 1996, p. 111).

Our study focuses on the two pollutants most common to the regulated facilities within our sample: biological oxygen demand (BOD) and total suspended solids (TSS). This focus maximizes the number of facilities for which data on wastewater limits and discharges are available. We examine each pollutant separately.

Both BOD and TSS are important pollutants. They are both conventional pollutants, they represent two of the most common pollutants, and they are discharged by many different industries. According to Kagan, Gunningham, and Thornton (2003, pp. 55–56), BOD, "the standard measure of organic pollutant content of water, is a universally important measure of effluent quality." According to Earnhart (2004c), because EPA considers BOD to be the most damaging of the conventional pollutants, BOD has been the focus of the agency's control efforts. According to one account, BOD is a measure of an emission's organic waste load that is "widely used as an index of harm from effluents, even though emissions that are low in BOD can still contain high levels of harmful inorganic pollutants" (Smith, 2000, p. 687). Kagan, Gunningham, and Thornton (2003, pp. 55–56) characterize TSS as "the standard measure of particulate content of water" and as "another universally important measure of effluent quality." In addition, the technologies used to control these pollutants appear to be the same or similar in many different industries. As a result, analysis of the limits imposed on discharges of these pollutants by the chemical industry are likely to be more generalizable than findings concerning other pollutants that are discharged by relatively few industries and that may require specialized control methods. Finally, although we lack specific information on this point, it is possible that the treatment technology used to comply with discharge limits for TSS or BOD may be effective in controlling other pollutants discharged by at least some industries. If so, our focus on these two pollutants should provide insights that may be transferable (1) to industries other than chemical manufacturing and (2) to efforts to control pollutants other than TSS and BOD expended by the chemical industry and other industries.

Facilities that discharge BOD and TSS rely on process modifications and treatment technologies to reduce discharge levels. Common process modifications might include a reduction in the amount of water used as an input in the manufacturing process, a reduction in the amount of wastewater generation, an increase in recycling, or a change in the identity of the chemicals used to facilitate removal of impurities or undesirable attributes of raw materials. As Chapter 5, Section 5.2 indicates, 88 percent of the facilities in our sample employ at least some form of TSS treatment. Of the technologies in use at those facilities, membrane filtration is the most effective. In addition, 81 percent of facilities in our sample employ at least some type of BOD treatment technology, with carbon adsorption being the most effective. The respondents in our survey indicated that their facilities' principal BOD removal technologies included solids removal, stabilization pond treatment, trickling filter biological treatment, activated sludge biological treatment, chemical oxidation, and carbon adsorption. The respondents' principal treatment technologies for TSS removal included coagulation plus sedimentation, sedimentation in effluent ponds, sedimentation in final clarifiers, granular media filtration, and membrane filtration. The EPA has relied on these same technologies in regulating discharges of BOD and TSS from other industries regulated under the Clean Water Act (Glicksman and Earnhart, 2007a). Further, these treatment technologies are not limited to the treatment of conventional pollutants, such as BOD and TSS. EPA has relied on some of the same treatment technologies to establish effluent limitation guidelines for toxic pollutants regulated under the Clean Water Act (Glicksman and Earnhart, 2007a).

Accordingly, we believe that the results of our analysis may be relevant beyond the narrow parameters of discharges of BOD and TSS by the chemical industry. If important elements relating to treatment technologies strongly determine regulatory stringency, and if the same treatment technologies help to control both TSS/BOD and other pollutants, then regulatory stringency may not differ much between TSS/BOD and these other pollutants. If they do not differ much, then an assessment of only TSS/BOD limits by extension sheds light on the regulatory stringency of the noted other pollutants. We address this point more fully in Section 4.6.

4.5. DATA SOURCE ON EFFLUENT LIMITS

The EPA maintains the Permit Compliance System (PCS) database as part of its implementation of the NPDES program. This database contains information on each and every discharge limit imposed on any regulated pollutant discharged by any permitted facility. This same database provides information on wastewater discharges. However, the database systematically records only limits imposed on and discharges released by major facilities.

As described in Chapter 2, this PCS database potentially provides data on discharge limits and wastewater discharges for 97 facilities that responded to our survey of chemical manufacturing facilities. The rest of our assessment focuses exclusively on these 97 facilities. In this chapter, we focus on the imposition of discharge limits on these facilities. As noted in Chapter 2, our sample period includes the 51 months between January 1999 and March 2003, inclusively.

4.6. IMPOSITION OF DISCHARGE LIMITS
IN CHOSEN SAMPLE

Before assessing the imposition of discharge limits, we first assess whether we should expect the imposition of discharge limits based on facilities' inclusion in the NPDES system. The PCS database identifies facilities as being "active" in the system or "inactive." Of the 97 facilities recorded in the PCS sample, nearly 89 percent of the facilities (86 of the 97) are active for the entire sample period (51 months). In contrast, 3 facilities are never active. Of the active facilities, one facility is active for only 4 of the 51 months and was inactive at the time of survey contact. Clearly, the EPA PCS database lists facilities as being contained within the NPDES system even when the facilities are inactive.

Our sample contains 4,947 facility-month observations (51 observations per each of the 97 facilities). Of these observations, 93 percent reflect an active month (4,604 months). With minor exceptions, we focus our analysis on the ever-active facilities (that is, facilities that are active throughout the sample period) and active months.

We begin our assessment of discharge limits by first considering the full range of regulated wastewater pollutants relevant for our respondent facilities. Later we focus on two important individual pollutants: TSS and BOD.

Of the 4,604 active months noted earlier, nearly 98 percent possess an operative limit for at least one regulated wastewater pollutant. After determining whether a facility is active in a particular month, we calculate the number of months with at least one operative limit for each separate facility. Based on this calculation, 72 of the 94 ever-active facilities possess at least one operative limit for each month in the sample period. In order to adjust more explicitly for the variation in active status per month, we calculate the ratio of months with operative limits to the number of active months for each separate facility. Eighty-five percent of the facilities possess at least one operative limit in each month that they are active.

Next we assess the presence of individual pollutants within our sample of respondent facilities. Based on the EPA PCS database, 336 individual wastewater pollutants are discharged by at least one of the respondent facilities. However, only 188 of these 336 pollutants are restricted by an operative limit for at least one of the respondent facilities. Of the 188 restricted pollutants, we identify the 15 most prevalent pollutants based on the number of facilities that at some point are restricted by the pollutant's discharge limit. We display these pollutants in increasing order of prevalence, with the count of relevant facilities indicated in parentheses: methylene chloride (12), phenol—single compound (12), pH range excursions—more than 60 minutes (13), pH range excursions—monthly total (13), cyanide (14), nickel (14), toluene (15), benzene (15), fecal coliform (17), oil and grease (19), chromium (21), copper (21), flow (21), chemical oxygen demand (23), zinc (23), nitrogen—ammonia (35), BOD (66), TSS (76).

This listing displays two pollutant categories that we ignore when identifying the 15 most prevalent pollutants. We disregard flow because it unto itself does not represent a pollutant; instead, this parameter is measured in order to eliminate the strategic option of diluting the presence of pollution in order to comply with a concentration-based discharge limit. We also disregard pH range excursions because this type of discharge limit is not comparable in format to any of the other limits. In particular, unlike the other discharge limits, these pH-related limits do not permit any assessment of overcompliance, which is prominent in our sample.

Far and away, the most prevalent regulated pollutant is TSS, and the second-most prevalent regulated pollutant is BOD. TSS and BOD affect 76 and 66 facilities, respectively. In comparison, the third-most prevalent pollutant affects only 35 facilities. At the other extreme, the fifteenth most prevalent pollutant, methylene chloride, affects only 12 facilities. Keep in mind that the sample includes 94 ever-active respondent facilities.[50]

Based on the top 15 pollutants listed previously, we demonstrate that our focus on TSS and BOD is not limiting since TSS limits, BOD limits, or both are related to limits imposed on other pollutants within the chemical manufacturing industry. We choose not to consider all 188 regulated pollutants since the number of facilities facing restrictions for the less prevalent pollutants is too small to warrant meaningful statistical assessment.

In the next two paragraphs, we describe the methodology we use to explore the relationships across the individual pollutant limits. Readers not skilled or interested in statistical analysis methodologies may wish to skip these paragraphs and focus on our description of the results later in the chapter. We calculate the Pearson correlation coefficients associated

with all of the relevant pairwise comparisons of the individual pollutants. Each correlation captures whether two pollutant limits vary together
in a systematic fashion. For our exploration, we first assess the signs of
these correlations in order to explore the types of relationships between
two particular pollutant limits. A positive correlation indicates that two
pollutant limits generally vary together in the same direction; that is, when
one limit is high, the other limit is high, and vice versa. A negative correlation indicates that two pollutant limits generally vary together in the
opposite direction; that is, when one limit is high, the other limit is low,
and vice versa. We also assess the magnitude of these correlations in order
to explore the strength of these relationships. Correlation magnitudes
above 0.50 generally indicate a reasonably strong relationship between
the pair of pollutant limits. In addition, we assess the statistical significance
of these correlations, as captured by the p-values associated with the correlations, in order to test whether the calculated magnitudes are distinguishable from zero; a zero magnitude implies no relationship between the two
particular pollutant limits. The calculated level of statistical significance,
reflected in the associated p-value, stems from a comparison of the correlation magnitude and the underlying variance of the two pollutant limits,
akin to a signal-to-noise ratio. We deem a correlation statistically significant when our analysis is able to reject the null hypothesis of a zero correlation at a significance level of 10 percent or less, indicating that the likelihood of incorrectly rejecting the null hypothesis lies at or below 10 percent.
This situation is revealed by a p-value of 0.10 or less. We deem a correlation statistically insignificant when our analysis is not able to reject the
null hypothesis of a zero correlation at a significance level of 10 percent or
less, indicating that the likelihood of incorrectly rejecting the null hypothesis lies above 10 percent. This situation is revealed by a p-value greater
than 0.10.

When assessing the correlations between pairwise comparisons of individual pollutants, we consider separately concentration-based limits
and quantity-based limits. We suspect a greater correlation between two
quantity-based limits than between two concentration-based limits since
the two limits in the first pair are both influenced by a common factor: a
facility's scale of operation (that is, production level). Moreover, for these
comparisons, we do not calculate the correlation based on facility-month
observations. Instead, we first calculate the median discharge-limit level
for each facility across the entire sample period. (A median represents
the value that divides a sample into two equally sized subsamples, half of
the sample lying below the median and half lying above the median.) While
this approach reduces the number of observations, it increases the likelihood that an operative limit applies to each considered pollutant in the

pairwise comparison, which is required for the correlation calculation. Obviously, this approach fails to compare the presence or absence of an operative limit across pairs of pollutants. We avoid this approach since the absence of an operative limit may not only reflect a fully lenient regulation but may also reflect the simple truth that a specific pollutant is not generated as a by-product of a particular production activity. Comparing the levels of operative limits avoids the latter possibility.

We assess first the correlations based on quantity-based limits. Consider TSS limits. Of the 13 other pollutants (which do not include BOD), 12 correlations exist in the sample (no facility faces operative quantity limits for both TSS and fecal coliform). Of these 12 correlations, 11 are positive. Of these 11 positive correlations, 8 are significantly positive and lie at or above 0.63. Only one correlation (copper) is negative, and it is insignificant so, that is, approximately zero. The correlations involving BOD limits reveal a similar pattern. Of the 13 other pollutants (which do not include TSS), 12 correlations exist in the sample (no facility faces operative quantity limits for both BOD and fecal coliform). Of these 12 correlations, 10 are positive. Of these 10 positive correlations, 8 are significantly positive and lie at or above 0.49. Only 2 correlations (copper and chromium) are negative and neither significantly so, that is, approximately zero. In general, these correlations reveal a sufficiently strong positive relationship between TSS and BOD limits and other prevalent pollutants, with the obvious exception of fecal coliform. The sample reveals no overlap with this exceptional pollutant; thus, we cannot claim to assess the influence of quantity-based fecal coliform limits on environmental behavior and performance. Fortunately, this absence of overlap should not affect our reported results since we exclude from our analyzed sample facilities that face neither TSS nor BOD limits.

We next assess the correlations based on concentration-based limits. While these correlations also support the conclusion of a positive relationship between TSS and BOD limits and other prevalent pollutants, the evidence is weaker. Consider TSS concentration limits. Of the 13 other pollutants, 10 correlations exist in the sample. Of these 10 correlations, 4 are positive. Of these 4 positive correlations, none are significantly positive, but all lie at or above 0.46. The lack of significance stems from the small sample sizes (the biggest sample includes 8 observations). In contrast, 4 correlations are negative; while none of these negative correlations is significantly different than zero, 2 are rather large, as the absolute magnitude lies at or above 0.45.

The correlations involving BOD concentration limits provide slightly stronger evidence. Of the 13 other pollutants, 12 correlations exist in the sample (no facility faces operative concentration limits for both BOD and fecal coliform). Of these 12 correlations, 9 are positive. Of these 9 positive correlations, 4 are significantly positive and lie at or above 0.90. An additional 4 correlations lie at or above 0.54. Three correlations prove negative. Of these 3 negative correlations, the correlation involving zinc is large ($\rho = -0.94$) and differs significantly from zero.[51] Another negative correlation lies at $\rho = -0.28$, which is meaningful but insignificantly different from zero.

In general, these concentration-limit correlations reveal a moderately positive relationship between TSS and BOD limits and other prevalent pollutants, with the obvious exceptions of fecal coliform and zinc. The absence of overlap with fecal coliform does not disrupt our reported results. However, our analysis of BOD concentration limits may be disrupted by the presence of a zinc limit. For those facilities where the zinc concentration limit proves the binding constraint on operations, our examination of BOD concentration limits may generate erroneous conclusions. This concern does not seem to apply to our examination of TSS concentration limits.

These points notwithstanding, we must acknowledge that the results of our examination of only TSS and BOD limits need not generalize to an examination of other pollutants.[52]

The discussed pairwise correlations indicate that TSS limits are sufficiently correlated with all but one of the other pollutant limits prevalent in our sample. A similar conclusion applies to BOD limits. Yet, we claim that we should not consider only one of these two key pollutants since the correlation between these two pollutants' limits is not sufficiently strong for both measurement forms. While the TSS and BOD quantity limits are extremely positively correlated ($\rho = .99$, $p = 0.0001$, $N = 59$), the TSS and BOD concentration limits are not strongly positively correlated ($\rho = 0.09$, $p = 0.7320$, $N = 16$).

4.7. IMPOSITION OF TSS LIMITS IN CHOSEN SAMPLE

In this section, we assess discharge limits imposed specifically on TSS; later in the chapter we assess discharge limits imposed specifically on BOD.

4.7.1. Presence of TSS Limits

Of the 94 ever-active respondent facilities, 83 facilities possess at least a single TSS-related record within the EPA PCS database. Based on a sample period of 51 months, these 83 facilities collectively generate a TSS subsample of 4,233 facility-month observations. Over the sample period, 76 facilities are always active, yet 2 facilities are inactive during the time period in which TSS-related records are present. Of the 4,233 observations, a facility is active in nearly 95 percent of the cases (4,020 observations).

Conditional upon active status, we calculate the presence of operative TSS limits. In general, we separately consider concentration limits and quantity limits. Of the 4,020 active facility-month observations, facilities face an operative TSS quantity limit in 3,269 cases, representing 81 percent of the TSS subsample. Considering each facility separately, we calculate the number of months with an operative TSS quantity limit, conditional on active status in a given month; of the 81 ever-active TSS-related facilities, 12 facilities never face an operative TSS quantity limit, while 50 facilities always face an operative TSS concentration limit over the 51 months of our study period.

We similarly consider operative TSS concentration limits. Of the 4,020 active facility-month observations, facilities face an operative TSS concentration limit in 1,273 cases, representing 32 percent of the TSS subsample. Considering each facility separately, we calculate the number of months with an operative TSS concentration limit, conditional on active status in a given month; of the 81 ever-active TSS-related facilities, 53 facilities never face an operative TSS concentration limit, while 16 facilities always face an operative TSS concentration limit.

Next we consider jointly concentration and quantity TSS limits. Of the 4,020 active facility-month observations, facilities face either an operative TSS quantity limit or concentration limit in 3,714 cases, representing 92 percent of the TSS subsample. Considering each facility separately, we calculate the number of months with an operative TSS concentration or quantity limit, or both, conditional on active status in a given month; of the 81 ever-active TSS-related facilities, 5 facilities never face an operative TSS limit, while 58 facilities always face an operative TSS limit during the 51 months of the study period.

Finally, we consider the combination of concentration and quantity TSS limits. Of the 4,020 active facility-month observations, facilities face both an operative TSS quantity limit and a concentration limit in 828 cases, representing 21 percent of the TSS subsample. In contrast, facilities face only an operative TSS quantity limit in 2,441 months (61 percent) and only an operative TSS concentration limit in 445 months (11 percent). Facilities face neither in 306 months (8 percent).

4.7.2. *Assessment of Operative Discharge Limit Levels*

Next we assess the levels of TSS discharge limits, not merely the presence of an operative limit. For this assessment, we employ two additional perspectives on the quantity and concentration limits. First, we anticipate that levels of quantity limits may be driven strongly by facility-specific production levels. In order to control for this variation in production so that we may compare effectively across facilities, we scale each facility's quantity limits by the facility's production level. For this task, we utilize information on production levels provided by facilities in response to a specific set of survey questions. The survey requested annual production levels for each year between 1999 and 2001, inclusively. (The first section of the appendix to Chapter 4 describes all of the survey questions analyzed in this book.) Unfortunately, only 73 of the 97 facilities provided production information for at least one of the three requested years. For those facilities reporting only a single year's data, we use that single data point to represent the facility's production level. For those facilities reporting data for two or three years, we use the average production level. For the remaining facilities, we impute facility-specific production levels based on sector-specific information on production-to-employee ratios or production without reference to employees. Of the remaining 24 facilities, 22 facilities provide information on their number of employees in response to a set of survey questions; specifically, the survey requested the annual number of employees for each year between 1999 and 2001, inclusively.[53] (The first section of the appendix to Chapter 4 describes all of the survey questions analyzed in this book.) Based on this information, we generate a proxy production level by multiplying the facility-specific number of employees by the relevant sector-specific median production-to-employee ratio. For this calculation, we consider both narrow sectors, based on the four-digit Standard Industrial Classification (SIC) code, and broad sectors: organic chemical manufacturing, inorganic chemical manufacturing, and other chemical manufacturing. When the relevant narrow sector includes data for at least 5 facilities, we utilize the narrow sector-specific median ratio (which applies to four narrow sectors: 2821, 2865, 2869, and 2873). Otherwise, we utilize the broad sector–specific median ratio. Of the 22 facilities lacking production data, 11 facilities utilize the narrow sector–specific median ratio, and 11 facilities utilize the broad sector–specific median ratio. Two facilities provide neither production nor employee information. For these facilities, we impute a production level based on the narrow sector–specific production median for one facility and broad sector–specific production median for the other facility.

Second, given the smaller sample of facilities facing TSS concentration limits, we attempt to expand the sample of facilities covered by limits that can be expressed in concentration terms. For this effort, we draw upon the quantity limits imposed on wastewater flow in order to convert TSS quantity limits into TSS concentration-equivalent limits based on the simple notion that a pollutant concentration equals the ratio of pollutant quantity to wastewater flow. We then merge these TSS concentration-equivalent limits with the TSS concentration limits (hereafter "concentration-equivalent"). In the process, we increase the sample size of operative TSS concentration-equivalent limits from 1,273 to 1,657.

Thus, we consider discharge limits in four forms: quantity limits, production-adjusted quantity limits, concentration limits, and concentration-equivalent limits. However, we focus most of our analysis on the production-adjusted quantity limits and the concentration-equivalent limits for two reasons. First, while we acknowledge that adjustments to production level may not fully adjust for regulatory differences across facilities, we cautiously claim that our assessment of production-adjusted quantity limits might reveal vast variation in discharge limits stemming from the application of state water-quality standards or best professional judgment on the notion that our adjustment by production sufficiently removes that dimension from consideration. Second, we wish to focus on the concentration-equivalent limits rather than the concentration limits since the sample is larger. We acknowledge that this broader consideration implies that our assessments of quantity limits and concentration-equivalent limits clearly overlap. Of course, an assessment of concentration limits would not remove this overlap since several facilities face both concentration and quantity limits.

For our assessment of discharge-limit levels, we must acknowledge that the data on discharge limits reported in the PCS database represent information on wastewater discharged from multiple pipes. Thus, multiple limit levels may apply to a specific facility in a given month. A substantial portion of the sample is influenced by the presence of multiple limit levels. Of the 3,269 relevant observations with operative quantity limits, 239 are influenced by multiple limit levels. Of the 1,272 relevant observations with operative concentration limits, 239 are influenced by multiple limit levels. Since government interventions—inspections and enforcement actions—apply to a particular facility, we do not wish to address individual pipes. We address the noted multiplicity of discharge limits by aggregating the data on limits to the combination of a particular facility and specific month within one year (hereafter "triplet"). We employ three aggregation protocols: mean or average within each triplet, median within each triplet, and minimum within each triplet.

Using pairwise correlations, we assess the differences across these three protocols. For the quantity limits, the three protocols generate nearly identical results. (The correlation between any of the two aggregated limits equals 0.999, p=0.0001.) Thus, it seems legitimate to use any one of these three aggregated limits. For the sake of comparability with the other three limit forms, we focus on the median- and minimum-aggregated limits. For the production-adjusted quantity limits, the median-aggregated limit is strongly correlated with both the minimum-aggregated limit ($\rho = 0.94$, $p = 0.0001$, $N = 3,269$) and the mean-aggregated limit ($\rho = 0.90$, $p = 0.0001$, $N = 3,269$). However, the minimum-aggregated limit is only moderately correlated with the mean-aggregated limit ($\rho = 0.69$, $p = 0.0001$, $N = 3,269$). Thus, we should assess the robustness of our results by exploring the minimum-aggregated limit and median- or mean-aggregated limit. Since the median protocol is less sensitive to outliers than is the mean protocol, we focus on the median protocol.

For the concentration limits, based on the correlations, no reasonable difference exists between the mean-aggregated limit and median-aggregated limit ($\rho = 0.999$, $p = 0.0001$); however, a substantial difference exists between these two aggregated limits and the minimum-aggregated limit (respectively; $\rho = 0.290$, $p = 0.0001$; $\rho = 0.283$, $p = 0.0001$). Thus, we retain at least the mean- or median-aggregated limit along with the minimum-aggregated limit; since the median protocol is less sensitive to outliers, we focus strongly on the median-aggregated limit. For the concentration-equivalent limits, the mean- and median-aggregated limits are nearly identical ($\rho = 0.9997$, $p = 0.0001$, $N = 1,657$). The minimum-aggregated limit is clearly different from the mean- and median-aggregated limits ($\rho = 0.74$, $p = 0.0001$, $N = 1,657$; $\rho = 0.73$, $p = 0.0001$, $N = 1,657$, respectively). Thus, we assess the robustness of our conclusions by considering both the minimum-aggregated limit and the median-aggregated limit.

We reassess the issue of discharge-limit level aggregation when examining the influence of limit levels on environmental behavior and wastewater discharges.

For the assessment of limit levels, we separately explore the distribution of quantity limits, production-adjusted quantity limits, concentration limits, and concentration-equivalent limits by reporting the average limit, median limit, and key percentiles (namely the twenty-fifth percentile and seventy-fifth percentile, which represent the first and third quartiles). First, we explore the distribution of quantity limit levels, as expressed in kilograms of TSS per day. As shown in Table 4.1, based on the median aggregation protocol, the interquartile range (first quartile to third quartile) lies between 81 kg/day and 445 kg/day; the average quantity limit equals 668 kg/day, while the median equals 239 kg/day. Based on the minimum aggregation protocol, the average quantity limit equals 649 kg/day. Second, we explore the distribution of production-adjusted quantity limit levels, as expressed in kilograms of TSS per 1,000 kilograms of product per day. As shown in Table 4.1, based on the median aggregation protocol, the interquartile range lies between 0.40 and 7.30 kg of TSS per 1,000 kg of product per day; the average limit equals 36.74, while the median equals 1.46.

TABLE 4.1

Discharge limits: Summary statistics

Table 4.1.a. TSS limits

Discharge Limit Form[a]	N	Mean	25th Percentile	Median	75th Percentile
Quantity Limits					
Median-Aggregated	3,269	666.650	80.649	239.497	444.974
Minimum-Aggregated	3,269	649.692	71.214	192.000	430.006
Concentration Limits					
Median-Aggregated	1,273	37.009	30.000	30.000	40.000
Minimum-Aggregated	1,273	31.777	25.000	30.000	30.000
Production-Adjusted Quantity Limits					
Median-Aggregated	3,269	36.742	0.403	1.463	7.304
Minimum-Aggregated	3,269	30.906	0.204	1.304	7.304
Concentration-Equivalent Limits					
Median-Aggregated	1,657	37.054	29.600	30.000	40.000
Minimum-Aggregated	1,657	32.934	25.000	30.000	31.697

Table 4.1.b. BOD limits

Discharge Limit Form[a]	N	Mean	25th Percentile	Median	75th Percentile
Quantity Limits					
Median-Aggregated	2,924	411.7	51.2	147.4	283.9
Minimum-Aggregated	2,924	400.1	43.6	113.4	280.3
Concentration Limits					
Median-Aggregated	860	26.5	17.9	24.0	30.0
Minimum-Aggregated	860	26.0	17.9	24.0	30.0
Production-Adjusted Quantity Limits					
Median-Aggregated	2,924	25.189	0.195	1.616	5.342
Minimum-Aggregated	2,924	25.061	0.164	1.211	5.198
Concentration-Equivalent Limits					
Median-Aggregated	1,181	24.974	14.000	20.776	30.000
Minimum-Aggregated	1,181	24.591	14.000	20.776	30.000

SOURCE: Environmental Protection Agency Permit Compliance System database.

[a]Multiple limits may be legally binding for a given facility in a specific moment in time (that is, year and month) due to the presence of multiple discharge pipes. These multiple limits are aggregated to the level of a specific facility and year-month combination by identifying the mean, median, or minimum value of the multiple limits.

Based on the minimum aggregation protocol, the average production-adjusted quantity limit equals 30.91 kg of TSS per 1,000 kg of product per day. Third, we explore the distribution of TSS concentration discharge-limit levels, as expressed in milligrams of TSS per liter. As shown in Table 4.1, based on the median aggregation protocol, the interquartile range of the concentration limit lies between 30 mg/L and 40 mg/L; the average limit level equals 37 mg/L, while the median operative limit equals 30 mg/L. Based on the minimum aggregation protocol, the average limit equals 32 mg/L. Fourth, we explore the distribution of TSS concentration-equivalent discharge-limit levels, as expressed in milligrams of TSS per liter. As shown in Table 4.1, based on the median aggregation protocol, the interquartile range lies between 29.6 and 40 mg/L, while the average limit level equals 37 mg/L and the median limit level equals 30 mg/L. Based on the minimum aggregation protocol, the average limit equals 32.9 mg/L.

4.7.3. Stringency of TSS Limits Across Various Subsamples

This section assesses the stringency of limits (that is, limit levels) across various subsamples within the TSS sample. (The second section of the appendix to this chapter assesses the presence of TSS limits across various subsamples.) For this assessment, we consider only production-adjusted quantity limits and concentration-equivalent limits.

4.7.3.1. Subsamples Based on Location—EPA Region. We first explore the variation in discharge limits across subsamples based on facility location as captured by EPA region. (We do not consider individual states because the number of facilities per state is small; thus, an assessment may reveal the identity of particular survey respondents.) We do not explore all seven regions represented in this TSS sample: Regions 1 through 7. Instead, we identify the two EPA regions with the largest number of TSS-related ever-active respondent facilities: Region 6 (with 28 facilities) and Region 4 (with 26 facilities). We then distinguish across Region 4, Region 6, and "Other" regions (with 27 facilities). First, we summarize the distribution of production-adjusted quantity limits by region based on the mean, median, and key percentiles: twenty-fifth percentile (first quartile) and seventy-fifth percentile (third quartile). As shown in Table 4.2, these summary statistics clearly indicate that limits vary across regions. Based on both the median- and minimum-aggregated limits, the regional median levels reveal that limits are tightest in Region 6 and loosest in "other" regions. Second, we summarize concentration-equivalent limits by region. As shown in Table 4.2, the region-specific medians are identical; thus, TSS concentration-equivalent limits do not differ across regions in the middle of the distributions. However, the distributions must differ at some points because the regional means differ. Based on the median-aggregated limits, the regional means reveal that concentration-equivalent limits are tightest in Region 4 and loosest in "other" regions.

TABLE 4.2

Discharge limits by EPA region: Summary statistics

Table 4.2.a. TSS limits

Discharge Limit Form[a]	Mean	25th Percentile	Median	75th Percentile
Production-Adjusted Quantity Limits				
Region "Other" (N=1,102)				
Median-Aggregated	47.079	0.792	5.342	20.059
Minimum-Aggregated	29.961	0.665	5.342	16.647
Region 4 (N=934)				
Median-Aggregated	53.567	0.572	1.463	5.559
Minimum-Aggregated	53.441	0.457	1.463	5.559
Region 6 (N=1,233)				
Median-Aggregated	14.759	0.147	0.897	5.816
Minimum-Aggregated	14.680	0.085	0.327	5.260
Concentration-Equivalent Limits				
Region "Other" (N=857)				
Median-Aggregated	38.197	29.600	30.000	40.000
Minimum-Aggregated	30.425	25.000	30.000	30.076
Region 4 (N=386)				
Median-Aggregated	34.806	30.000	30.000	50.000
Minimum-Aggregated	34.806	30.000	30.000	50.000
Region 6 (N=414)				
Median-Aggregated	36.785	21.379	30.000	31.697
Minimum-Aggregated	36.380	21.379	30.000	31.697

SOURCE: Environmental Protection Agency Permit Compliance System database.

[a]Multiple limits may be legally binding for a given facility in a specific moment in time (that is, year and month) due to the presence of multiple discharge pipes. These multiple limits are aggregated to the level of a specific facility and year-month combination by identifying the mean, median, or minimum value of the multiple limits.

4.7.3.2. Subsamples Based on Chemical Manufacturing Subsectors. Next we assess the stringency of limits (that is, limit levels) across subsamples based on chemical manufacturing subsectors: organic chemical manufacturing, inorganic chemical manufacturing, and other chemical manufacturing. First, we assess production-adjusted quantity limits by sector subsample. As shown in Table 4.3, considering the median-aggregated limits, facilities manufacturing inorganic chemicals clearly face the most stringent limits, based on the median and quartile values. However, based on the mean values, these facilities face the least stringent limits. Based on both the mean and median values, organic chemical manufacturing facilities face limits tighter than those faced by facilities manufacturing

TABLE 4.2 *(continued)*

Table 4.2.b. BOD limits

Discharge Limit Form[a]	Mean	25th Percentile	Median	75th Percentile
Production-Adjusted Quantity Limits				
Region "Other" (N = 1057)				
Median-Aggregated	33.037	0.443	3.592	20.318
Minimum-Aggregated	32.876	0.443	3.592	20.318
Region 4 (N = 867)				
Median-Aggregated	34.467	0.265	0.877	3.001
Minimum-Aggregated	34.287	0.195	0.877	2.947
Region 6 (N = 1,000)				
Median-Aggregated	8.848	0.132	0.600	3.058
Minimum-Aggregated	8.802	0.097	0.383	3.058
Concentration-Equivalent Limits				
Region "Other" (N = 792)				
Median-Aggregated	29.685	20.000	26.000	31.200
Minimum-Aggregated	29.170	20.000	26.000	31.200
Region 4 (N = 165)				
Median-Aggregated	17.522	4.000	22.300	30.000
Minimum-Aggregated	17.522	4.000	22.300	30.000
Region 6 (N = 224)				
Median-Aggregated	13.809	6.892	12.509	19.977
Minimum-Aggregated	13.609	6.892	12.509	18.728

SOURCE: Environmental Protection Agency Permit Compliance System database.

[a]Multiple limits may be legally binding for a given facility in a specific moment in time (that is, year and month) due to the presence of multiple discharge pipes. These multiple limits are aggregated to the level of a specific facility and year-month combination by identifying the mean, median, or minimum value of the multiple limits.

"other" chemicals. Use of the minimum-aggregated limits reveals no discrepancy between the mean-based ranking and that median-based ranking; regardless of the summary statistic, inorganic chemical facilities face the tightest limits, while "other" chemical facilities face the least stringent limits.

Second, we assess concentration-equivalent limits. As shown in Table 4.3, across the full distribution, limits are looser for facilities manufacturing "other" chemicals, based on either median- or minimum-aggregated limits. Moreover, organic chemical facilities face limits tighter than (or equal to) those faced by inorganic chemical facilities, based on median-aggregated limits. An assessment of minimum-aggregated limits generates the same general conclusion.

TABLE 4.3

Discharge limits by chemical manufacturing subsector: Summary statistics

Table 4.3.a. TSS limits

Discharge Limit Form[a]	Mean	25th Percentile	Median	75th Percentile
Production-Adjusted Quantity Limits				
Inorganic Chemical Subsector (N = 652)				
Median-Aggregated	55.765	0.175	0.897	3.436
Minimum-Aggregated	27.917	0.175	0.897	3.436
Organic Chemical Subsector (N = 2,133)				
Median-Aggregated	31.197	0.588	1.639	11.862
Minimum-Aggregated	30.765	0.204	1.481	11.862
"Other" Chemical Subsector (N = 484)				
Median-Aggregated	35.554	0.457	2.969	80.188
Minimum-Aggregated	35.554	0.457	2.969	80.188
Concentration-Equivalent Limits				
Inorganic Chemical Subsector (N = 532)				
Median-Aggregated	39.551	30.000	30.000	50.000
Minimum-Aggregated	30.923	25.000	30.000	30.076
Organic Chemical Subsector (N = 899)				
Median-Aggregated	32.436	27.500	30.000	32.213
Minimum-Aggregated	29.947	25.000	29.600	31.697
"Other" Chemical Subsector (N = 226)				
Median-Aggregated	49.547	30.000	38.217	40.000
Minimum-Aggregated	49.547	30.000	38.217	40.000

SOURCE: Environmental Protection Agency Permit Compliance System database.

[a]Multiple limits may be legally binding for a given facility in a specific moment in time (that is, year and month) due to the presence of multiple discharge pipes. These multiple limits are aggregated to the level of a specific facility and year-month combination by identifying the mean, median, or minimum value of the multiple limits.

4.7.3.3. Subsamples Based on Facility Size. Next we assess the stringency of limits across subsamples based on facility size. Of course, various measures may capture facility size. For our purposes, we measure facility size based on the number of facility-level employees measured in full-time equivalent terms. Our survey gathers year-specific values on the number of facility-level employees for the years 1999, 2000, and 2001. Rather then considering year-specific values, we identify the average of these year-specific values as the one measure of facility size.[54] We assess the connection between facility size and limits in two ways. First, we divide our sample into two subsamples based on the sample median value of facility size, which equals 286.7. Then we compare below-median-size facilities and above-median-size facilities. Next, we explore the variation in discharge

TABLE 4.3 *(continued)*

Table 4.3.b. BOD limits

Discharge Limit Form[a]	Mean	25th Percentile	Median	75th Percentile
Production-Adjusted Quantity Limits				
Inorganic Chemical Subsector (N=356)				
Median-Aggregated	34.893	0.195	0.600	2.750
Minimum-Aggregated	34.893	0.195	0.600	2.750
Organic Chemical Subsector (N=2,040)				
Median-Aggregated	20.102	0.289	1.616	6.422
Minimum-Aggregated	19.919	0.164	1.211	5.198
"Other" Chemical Subsector (N=528)				
Median-Aggregated	38.298	0.062	2.485	66.363
Minimum-Aggregated	38.298	0.062	2.485	66.363
Concentration-Equivalent Limits				
Inorganic Chemical Subsector (N=298)				
Median-Aggregated	21.435	8.000	30.000	30.000
Minimum-Aggregated	21.435	8.000	30.000	30.000
Organic Chemical Subsector (N=659)				
Median-Aggregated	21.292	14.084	20.000	24.000
Minimum-Aggregated	20.605	14.084	20.000	24.000
"Other" Chemical Subsector (N=224)				
Median-Aggregated	40.516	29.184	40.078	53.000
Minimum-Aggregated	40.516	29.184	40.078	53.000

SOURCE: Environmental Protection Agency Permit Compliance System database.

[a]Multiple limits may be legally binding for a given facility in a specific moment in time (that is, year and month) due to the presence of multiple discharge pipes. These multiple limits are aggregated to the level of a specific facility and year-month combination by identifying the mean, median, or minimum value of the multiple limits.

limits and its connection to the facility-specific number of employees without any reference to the sample median number of employees. For this approach, we assess the correlation between discharge-limit levels and number of employees.

Initially, we assess the distribution of limit levels by facility size category based on mean, median, and quartile values. We first summarize the production-adjusted quantity limits. As shown in Table 4.4, these limits differ between the smaller and larger facilities but by a surprisingly moderate degree. We also summarize the concentration-equivalent limits. As shown in Table 4.4, the two subsamples of facilities appear to face highly similar limits.

In addition to exploring differences between subsamples based on facility size, we explore the correlation between discharge-limit levels and

TABLE 4.4

Discharge limits by facility size category: Summary statistics

Table 4.4.a. TSS limits

Discharge Limit Form[a]	Mean	25th Percentile	Median	75th Percentile
Production-Adjusted Quantity Limits				
Below-Median Size (N = 1,479)				
Median-Aggregated	27.830	0.376	1.613	7.002
Minimum-Aggregated	27.350	0.277	1.462	7.002
Above-Median Size (N = 1,790)				
Median-Aggregated	44.105	0.403	1.304	12.024
Minimum-Aggregated	33.844	0.147	1.276	12.024
Concentration-Equivalent Limits				
Below-Median Size (N = 860)				
Median-Aggregated	34.459	29.600	30.000	31.697
Minimum-Aggregated	33.436	25.000	30.000	30.000
Above-Median Size (N = 797)				
Median-Aggregated	39.855	29.450	30.000	40.000
Minimum-Aggregated	32.392	23.400	30.000	38.217

Table 4.4.b. BOD limits

Discharge Limit Form[a]	Mean	25th Percentile	Median	75th Percentile
Production-Adjusted Quantity Limits				
Below-Median Size (N = 1,451)				
Median-Aggregated	18.204	0.383	1.911	5.198
Minimum-Aggregated	18.087	0.383	1.911	4.673
Above-Median Size (N = 1,473)				
Median-Aggregated	32.069	0.114	1.211	20.318
Minimum-Aggregated	31.931	0.062	0.782	20.318
Concentration-Equivalent Limits				
Below-Median Size (N = 593)				
Median-Aggregated	21.638	12.509	20.000	26.000
Minimum-Aggregated	21.638	12.509	20.000	26.000
Above-Median Size (N = 588)				
Median-Aggregated	28.339	14.084	24.000	30.638
Minimum-Aggregated	27.569	14.084	24.000	30.638

SOURCE: Environmental Protection Agency Permit Compliance System database.

[a]Multiple limits may be legally binding for a given facility in a specific moment in time (that is, year and month) due to the presence of multiple discharge pipes. These multiple limits are aggregated to the level of a specific facility and year-month combination by identifying the mean, median, or minimum value of the multiple limits.

facility size. Consider first the correlation between the facility-specific number of employees and the facility-specific production-adjusted quantity limit. Bigger facilities face looser limits, at least based on the median-aggregated limit ($\rho = -0.49$, $p = 0.005$, $N = 3{,}269$). However, no significant correlation exists based on the minimum-aggregated limit ($\rho = -0.003$, $p = 0.87$, $N = 3{,}269$). Consequently, our conclusion is not robust to the aggregation protocol. Consider next the correlation between a facility-specific number of employees and the facility-specific concentration-equivalent limit. In this case, bigger facilities face *tighter* limits. The correlation coefficient is significant based on the minimum-aggregated limit ($\rho = -0.06$, $p = 0.01$, $N = 1{,}657$) but not on the median-aggregated limit ($\rho = -0.03$, $p = 0.21$, $N = 1{,}657$). Thus, our conclusion is regrettably not robust to the choice of aggregation protocol. As important, our conclusion regarding discharge limits in general is not robust to the form of the discharge limit.

4.7.3.4. Subsamples Based on Time: Years 1999–2003   We next assess the stringency of limits across subsamples based on time, as divided across the five years represented in our sample (1999 through 2003). First, we explore production-adjusted quantity limits across time. As shown in Table 4.5, the median values are trending down over time. In contrast, the mean values are higher in 2002 and 2003 relative to the earlier years. Second, we explore concentration-equivalent limits. As shown in Table 4.5, the median value does not vary at all over time. However, the mean value is clearly higher in 2002 and 2003 relative to the earlier years. (Table 4.5 also reports the first and third quartile values by year.)

4.7.4. Assessment of Variation in Presence and Stringency of Limits Across Time for Given Facilities

Finally, our exploration of discharge limits assesses variation in the presence and stringency of limits across time for given facilities. We separately assess quantity limits and concentration limits. (Separate consideration of production-adjusted quantity limits is not needed since production, by design, does not vary over time; separate consideration of concentration equivalent is not warranted since the assessment considers the pattern for each individual facility so sample size is much less important.) Of the 81 ever-active, TSS-relevant facilities, 69 facilities at some point face an operative quantity limit. Of these 69 facilities, the median-aggregated limit level varies over time for 25 facilities, and the minimum-aggregated limit varies over time for 24 facilities; thus, 44 to 45 facilities (64 to 65 percent of facilities) face the same operative discharge limit over the entire sample period.

TABLE 4.5

Discharge limits by sample year: Summary statistics

Table 4.5.a. TSS limits

Year	Discharge Limit Form[a]	Mean	25th Percentile	Median	75th Percentile
	Production-Adjusted Quantity Limits				
1999 (N=735)	Median-Aggregated	36.600	0.457	1.481	7.304
	Minimum-Aggregated	30.988	0.277	1.463	7.304
2000 (N=776)	Median-Aggregated	34.767	0.403	1.463	7.304
	Minimum-Aggregated	29.458	0.204	1.304	7.304
2001 (N=789)	Median-Aggregated	35.495	0.403	1.481	7.304
	Minimum-Aggregated	30.081	0.199	1.304	7.304
2002 (N=778)	Median-Aggregated	39.283	0.376	1.463	7.462
	Minimum-Aggregated	32.536	0.199	1.276	7.462
2003 (N=191)	Median-Aggregated	40.114	0.277	1.304	7.462
	Minimum-Aggregated	33.242	0.175	1.203	7.462
	Concentration-Equivalent Limits				
1999 (N=280)	Median-Aggregated	36.030	30.000	30.000	40.000
	Minimum-Aggregated	30.826	25.000	30.000	30.076
2000 (N=386)	Median-Aggregated	35.080	27.500	30.000	39.017
	Minimum-Aggregated	30.915	25.000	30.000	31.697
2001 (N=421)	Median-Aggregated	36.708	29.600	30.000	40.000
	Minimum-Aggregated	32.738	25.000	30.000	34.360
2002 (N=456)	Median-Aggregated	39.239	29.600	30.000	40.000
	Minimum-Aggregated	35.577	25.000	30.000	34.360
2003 (N=114)	Median-Aggregated	38.796	29.600	30.000	40.000
	Minimum-Aggregated	35.096	25.000	30.000	32.213

SOURCE: Environmental Protection Agency Permit Compliance System database.

[a]Multiple limits may be legally binding for a given facility in a specific moment in time (that is, year and month) due to the presence of multiple discharge pipes. These multiple limits are aggregated to the level of a specific facility and year-month combination by identifying the mean, median, or minimum value of the multiple limits.

Of the 81 ever-active, TSS-relevant facilities, 28 facilities at some point face an operative concentration limit. Of these 28 facilities, the median-aggregated limit level varies over time only for a single facility, and the minimum-aggregated limit never varies for all facilities.

However, this representation masks the presence of switching between an operative limit and a nonoperative limit. For this further assessment, we consider all of the TSS-relevant facilities, including the two facilities that were never active during our sample period. Of these 83 facilities, 44 facilities face the same quantity limit throughout the sample period (including the 2 never-active facilities). Another 33 facilities face two different quantity limit levels; 10 of these facilities face two quantity limits merely because at some point they do not face an operative limit. The other 23 fa-

TABLE 4.5 *(continued)*

Table 4.5.b. BOD limits

Year	Discharge Limit Form[a]	Mean	25th Percentile	Median	75th Percentile
	Production-Adjusted Quantity Limits				
1999 (N=668)	Median-Aggregated	25.730	0.289	1.895	5.618
	Minimum-Aggregated	25.655	0.195	1.895	5.342
2000 (N=699)	Median-Aggregated	24.895	0.195	1.211	5.342
	Minimum-Aggregated	24.823	0.176	1.211	5.198
2001 (N=704)	Median-Aggregated	24.639	0.176	1.211	5.342
	Minimum-Aggregated	24.495	0.156	1.211	4.673
2002 (N=683)	Median-Aggregated	25.459	0.176	1.636	6.422
	Minimum-Aggregated	25.259	0.132	1.211	5.198
2003 (N=170)	Median-Aggregated	25.458	0.176	1.636	6.422
	Minimum-Aggregated	25.257	0.132	0.877	5.198
	Concentration-Equivalent Limits				
1999 (N=213)	Median-Aggregated	25.012	17.900	24.000	30.000
	Minimum-Aggregated	24.561	17.900	24.000	30.000
2000 (N=283)	Median-Aggregated	24.372	14.000	20.776	30.000
	Minimum-Aggregated	23.993	12.509	20.776	30.000
2001 (N=295)	Median-Aggregated	25.253	14.000	20.776	30.000
	Minimum-Aggregated	24.877	14.000	20.776	30.000
2002 (N=312)	Median-Aggregated	25.218	14.084	20.903	30.000
	Minimum-Aggregated	24.862	14.084	20.903	30.000
2003 (N=78)	Median-Aggregated	25.027	14.084	20.903	30.000
	Minimum-Aggregated	24.672	14.084	20.903	30.000

SOURCE: Environmental Protection Agency Permit Compliance System database.

[a]Multiple limits may be legally binding for a given facility in a specific moment in time (that is, year and month) due to the presence of multiple discharge pipes. These multiple limits are aggregated to the level of a specific facility and year-month combination by identifying the mean, median, or minimum value of the multiple limits.

cilities face two operative quantity limits. Four facilities face three different quantity limits. Only 1 of these facilities faces three different operative limit levels. The other 3 facilities face two different operative limits and a nonoperative limit at some point. Finally, 2 facilities face four different limit levels. One of these facilities faces four different operative limit levels. The other faces three different operative limit levels and a nonoperative limit at some point.

Next we assess concentration limit levels. Of the 83 TSS-relevant facilities, 70 facilities face the same concentration limit throughout the sample period. However, 13 facilities face two concentration limit levels; for 12 of these facilities, this variation stems from the presence of both an operative limit and a nonoperative limit.

Finally, we identify each facility's most stringent limit (smallest limit level across the sample period, based on the minimum-aggregated limit) and then assess the prevalence of observations that represent deviations from this facility-specific minimum limit level. Data on quantity limits reveal that 81.5 percent of the observations do not deviate from the facility-specific minima. Clearly, a substantial portion of the sample reflects a quantity limit level that is constant across the entire sample period. Data on concentration limits show no deviations.

4.8. IMPOSITION OF BOD LIMITS
IN CHOSEN SAMPLE

In this final section of Chapter 4, we assess discharge limits imposed specifically on BOD.

4.8.1. *Presence of BOD Limits*

Of the 94 ever-active respondent facilities, 68 facilities possess at least a single BOD-related record within the EPA PCS database. Based on a sample period of 51 months, these 68 facilities collectively generate a BOD subsample of 3,468 facility-month observations. Over the sample period, nearly all (65) are always active, and all facilities are active at some point. Of the 3,468 observations, a facility is active in 97.5 percent of the cases (3,380 observations).

Conditional upon active status, we calculate the presence of operative BOD limits. In general, we separately consider concentration limits and quantity limits. Of the 3,380 active facility-month observations, facilities face operative BOD quantity limits in 87 percent of the sample (2,924 observations) and face operative BOD concentration limits in 25 percent of the sample (860 observations). Facilities face either form of a BOD limit in 92 percent of the sample (3,117 observations). Lastly, facilities face both forms of a BOD limit in 20 percent of the sample (667 observations).

Considering each facility separately, we calculate the number of months with an operative BOD limit, conditional on active status in a given month. Some facilities never face an operative quantity limit (5 facilities, 7 percent of sample), while other facilities always face an operative quantity limit (42 facilities, 62 percent of sample). Some facilities never face an operative concentration limit (45 facilities, 66.2 percent of sample), while other facilities always face an operative concentration limit (8 facilities, 12 percent of sample). Some facilities never face an operative limit (2 facilities, 3 percent of sample), while other facilities always face an operative limit (47 facilities, 69 percent of sample).

4.8.2. Assessment of Operative Discharge Limit Levels

Next we assess the levels of BOD discharge limits. As with TSS, we consider BOD discharge limits in four forms: quantity limits, production-adjusted quantity limits, concentration limits, and concentration-equivalent limits, with a stronger focus on production-adjusted quantity limits and concentration-equivalent limits. And as with TSS, we employ three aggregation protocols for measuring the discharge-limit level for an individual facility within a given month and year: mean aggregation, median aggregation, and minimum aggregation. Correlations demonstrate that the three protocols generate nearly identical results.[55] Thus, it is clearly legitimate to use any one of the three aggregated limits. For the sake of comparability with the TSS limits, we tabulate both the median- and minimum-aggregated limits; however, we assess only the latter form. Regardless, we consider both forms of aggregation when examining the influence of BOD limit levels on environmental behavior and wastewater discharges.

For the assessment of limit levels, we separately explore the distribution of quantity limits, production-adjusted quantity limits, concentration limits, and concentration-equivalent limits by reporting the average limit, the median limit, and the first and third quartiles. First, we explore the distribution of quantity limit levels, as expressed in kilograms of BOD per day. As shown in Table 4.1.b, based on the median aggregation protocol, the interquartile range lies between 51 kg/day and 284 kg/day; in between, the median limit sits at 147 kg/day. Based on the minimum aggregation protocol, the median quantity limit equals 113 kg/day. Second, we explore the distribution of production-adjusted quantity limit levels, as expressed in kilograms of BOD per 1,000 kilograms of product per day. As shown in Table 4.1.b, based on the median aggregation protocol, the interquartile range lies between 0.19 and 5.34 kg of BOD per 1,000 kg of product per day; in between, the median limit sits at 1.62 kg of BOD per 1,000 kg of product per day. Based on the minimum aggregation protocol, the median production-adjusted quantity limit equals 1.21 kg of BOD per 1,000 kg of product per day. Third, we explore the distribution of BOD concentration discharge-limit levels, as expressed in milligrams of TSS per liter. As shown in Table 4.1.b, regardless of the aggregation protocol, the interquartile range of the concentration limit lies between 18 mg/L and 30 mg/L; in between, the median limit sits at 24 mg/L. Fourth, we explore the distribution of TSS concentration-equivalent discharge-limit levels, as expressed in milligrams of TSS per liter. As shown in Table 4.1.b, regardless of the aggregation protocol, the interquartile range lies between 14 and 30 mg/L; in between, the median limit sits at 20.8 mg/L.

4.8.3. Stringency of BOD Limits Across
Various Subsamples

This section assesses the stringency of limits across various subsamples within the BOD sample. (The second section of the appendix to this chapter assesses the presence of BOD limits across various subsamples.) For this assessment, we consider only production-adjusted quantity limits and concentration-equivalent limits.

4.8.3.1. Subsamples Based on Location—EPA Region. We first explore the variation in discharge limits across subsamples based on facility location as captured by EPA region. Again, Regions 4 and 6 prove the most prominent regions. Region 6 houses 22 BOD-relevant, ever-active facilities, and Region 4 houses 21 facilities. For this exploration, we first summarize the distribution of production-adjusted quantity limits by region based on the mean, median, and first quartile and third quartile values. As shown in Table 4.2.b, facilities operating in Regions 4 and 6 face tighter limits than facilities operating in "other" regions. Similarly, we assess variation in concentration-equivalent limits across regions. Unfortunately, as shown in Table 4.2.b, most of the observations fall outside of the regions chosen for close examination. Consequently, we cautiously report that facilities appear to face tighter limits in Region 6 than in "other" regions.

4.8.3.2. Subsamples Based on Chemical Manufacturing Subsectors. Next we assess the stringency of limits across subsamples based on chemical manufacturing subsectors: organic chemical manufacturing, inorganic chemical manufacturing, and other chemical manufacturing. First, we assess production-adjusted quantity limits by sector subsample. As shown in Table 4.3.b, noticeable differences in BOD production-adjusted quantity limit levels exist across the sectors; in particular, facilities manufacturing organic or inorganic chemicals face tighter limits than do facilities manufacturing "other" chemicals. Similarly, noticeable differences in BOD concentration-equivalent limit levels exist across the sectors; as with production-adjusted quantity limits, facilities manufacturing organic or inorganic chemicals face tighter concentration-equivalent limits than do facilities manufacturing "other" chemicals.

4.8.3.3. Subsamples Based on Facility Size. Next we assess the stringency of limits across subsamples based on facility size. Again, we use the average number of employees working over the three-year period of 1999 to 2001 to measure facility size.[56]

As with TSS limits, we assess the connection between facility size and BOD limits in two ways. We divide our sample into two subsamples based on the sample median value of facility size, which equals 348. Then we compare below-median-size facilities and above-median-size facilities. We also explore the correlation between facility-specific discharge-limit levels and number of employees without any reference to the median-size facility.

Initially we assess the distribution of limit levels by facility size category based on mean, median, and quartile values. We first summarize the production-adjusted quantity limits. As shown in Table 4.4.b, larger facilities face tighter limits, based on the median values, but face looser limits, based on the mean values. The distinction appears driven by the presence of much looser limits imposed on larger facilities in the upper end of the distribution but tighter limits imposed on larger facilities in the lower end of the distribution, with the former effect dominating. We also assess concentration-equivalent limits. As shown in Table 4.4.b, it appears that larger facilities face looser limits.

In addition to splitting the sample according to the sample median facility size, we explore the correlation between the actual number of employees employed by a specific facility and the facility-specific limit level. Based on an assessment of production-adjusted quantity limits, we conclude that larger facilities face looser limits, even after adjusting for production, regardless of the aggregation protocol. Based on the median-aggregated limit and the minimum-aggregated limit, the correlation is significantly positive ($\rho = 0.055$, $p = 0.003$, $N = 2,924$, in both cases). In contrast, based on an assessment of concentration-equivalent limits, larger facilities face tighter concentration limits, though the correlation is small, regardless of the aggregation protocol. The correlation proves significant only based on the minimum aggregation protocol. (Based on the median-aggregated limit and the minimum-aggregated limit, the correlation is negative ($\rho = -0.029$, $p = 0.32$; $\rho = -0.074$, $p = 0.01$, respectively; $N = 1,181$).

In sum, our conclusion is not robust to the form of discharge limit analyzed and the analytical approach. Larger facilities may face more or less stringent limits than those faced by smaller facilities.

4.8.3.4. Subsamples Based on Time: Years 1999—2003. We next assess the stringency of limits across subsamples based on time, as divided across the five years represented in our sample (1999 through 2003). We assess the levels of operative limits by exploring mean and median values. As shown in Table 4.5.b, production-adjusted quantity limit levels trended down between 1999 and 2000 and between 2002 and 2003. BOD concentration-equivalent limit levels dropped after 1999 but then leveled off, based on the median values; the mean values do not differ to any meaningful degree over the five-year period. (Table 4.5.b also reports the first and third quartile values by year.)

4.8.4. Assessment of Variation in Presence and Stringency
of Limits Across Time for Given Facilities

Finally, our exploration of discharge limits assesses variation in the presence and stringency of limits across time for given facilities. As with TSS limits, we assess only BOD quantity limits and concentration limits. In the case of concentration limits, no variation over time exists. If a facility ever faces an operative limit, the level is constant over time. In the case of quantity limits, 42 facilities (67 percent of the sample ever facing quantity limits) face a single operative quantity limit level over the sample period. In contrast, 21 facilities face either two or three different operative limit levels. (Five facilities never face operative quantity limits.) As an alternative angle, we assess the number of observations where the operative limit represents a deviation from the relevant facility's sample-period minimum limit level. Of the 2,924 observations in which a facility faces an operative quantity limit, 83 percent of the observations reveal no deviation in the limit level relative to the facility's own minimum level. In sum, the sample reveals little variation in the discharge limit imposed on an individual facility over the five-year sample period.

4.9. SUMMARY

By describing the structure of the Clean Water Act, this chapter provides the reader with background information that provides context for the empirical assessments described in Chapters 5 though 9. It is important to understand the roles of EPA and the states under the Clean Water Act, the manner in which EPA sets effluent limitation guidelines that are later used (principally by the states) to craft discharge limits for individual facilities, the operation of the NPDES permit program, the role of state water-quality standards in the permit process, and the reasons why point sources in the same industrial category may be subject to widely disparate discharge limits. The chapter also describes the database that serves as the source of information for our study of discharge limits and discharges of TSS and BOD by facilities in the chemical industry. These limits are correlated with limits imposed on other pollutants discharged by the chemical industry. Our focus here is on production-adjusted quantity limits and concentration-equivalent limits, as we assess the presence and stringency of discharge limits across regions, across industrial subsectors, and over time.

Chapter Four Appendix

This first section of the appendix to Chapter 4 lists the relevant questions posed in our survey of chemical manufacturing facilities. We display only those questions analyzed in this book. Moreover, we display the questions in the order in which they appear in the book, not as they appeared in the survey instrument.

1. What was your facility's total annual product output in Year X (in tons)?
2. About how many full-time-equivalent employees did your facility have in Year X?
3. In Year X, about how many person-months of time did your facility allocate to help ensure that the facility met environmental regulations? For the purposes of this survey, we will call these people environmental employees, and they may include contract employees, consultants, engineers, operators, and maintenance personnel.
4. About how many employees work on wastewater in your facility?
5. About how many full-time-equivalent employees did your facility have in Year X?
6. About what percent of the environmental employees at your facility hold a four-year college education (e.g., BS or BA)?
7. On average, about how many years of experience does the typical environmental employee who works at your facility have?
8. Does your facility currently provide internal or external training programs for your environmental employees to ensure compliance with environmental regulations related to water discharges?
9. About how many days of training are provided to employees each year?

10. About what percent of your environmental employees attend this training?
11. What is your principal TSS removal technology prior to effluent discharge?
    – No Treatment
    – Coagulation plus Sedimentation
    – Sedimentation in Effluent Pond
    – Sedimentation in Final Clarifier
    – Granular Media Filtration
    – Membrane Filtration
    – Other
12. What is your principal BOD removal technology prior to effluent discharge?
    – No Treatment
    – Solids Removal
    – Stabilization Pond Treatment
    – Trickling Filter Biological Treatment
    – Activated Sludge Biological Treatment
    – Chemical Oxidation
    – Carbon Adsorption
    – Other
13. Did your facility complete at least one upgrade of its wastewater treatment facilities in the past 3 years?
14. In what year(s) did you complete upgrades?
15. Were any of the facility upgrades designed to reduce directly or indirectly the concentration of pollutants discharged to surface water?
16. Of the following, which best describes your wastewater monitoring program?
    – No Monitoring (except final discharge)
    – Treatment Process Monitoring
    – Single-Point Sewer and Treatment Process
    – Multiple-Point Sewer and Treatment Process
17. In Year X, about how many times did your facility conduct environmental self-audits that included NPDES discharge operations?
18. If at all, who generally conducts environmental self-audits at your facility?
    – No Internal Audits
    – Visiting Corporate and/or Consultants Only
    – Teams of Various Facility Employees
    – Teams of Facility Employees and Visiting Corporate Staff and/ or Consultants

19. What best describes your facility's protocols for classifying self-audit findings and observations?
    – No Audits
    – Only Noncompliance Findings Noted
    – Two Categories Used: Noncompliance vs Other
    – Findings Placed into Priority Categories of Vulnerability for Noncompliance
    – Findings Ranked by Degree of Vulnerability for Noncompliance
20. Over the past 3 years, has your facility used external consultants to address wastewater issues at your facility?
21. Over the past 3 years, approximately how much does your facility spend on external wastewater consultants in an average year?
22. Over the past 3 years, did your facility seek assistance from your government water regulator on ways in which your facility could reduce or further reduce its wastewater discharge below the permitted discharge limit?
23. Over the past 3 years, how frequently did environmental compliance assistance resources meet the day-to-day needs of your facility's environmental employees (e.g., internal resources, trade association resources, governmental assistance, EPA hotline)?
    – Never
    – Some of the Time
    – Most of the Time
    – Always
24. What level of importance does the facility's management place on the role of general plant employees (i.e., all nonmanagement employees) in identifying and correcting conditions that may lead to noncompliance with environmental regulations for water pollution discharges?
    – Not Important
    – Somewhat Important
    – Important
    – Very Important
25. Of the following, what steps does the facility's management take to communicate water pollutant discharge performance goals and targets to facility employees?
    – Regularly Scheduled Meetings or Briefings
    – Regular Memorandum, Newsletter, or Email
    – Public Display Board
    – General Plant Worker Training or Orientation Session
    – As a Review Item for New Design or Construction Projects
    – Other

26. Of the following, what steps does the facility's management take to communicate environmental performance goals and targets to external parties, such as community leaders, the mass media, etc.?
    – Press releases
    – Advertising
    – Open House
    – Focus Groups
    – Web sites or Internet Messages
    – Other
27. Is your facility currently compliant with ISO 14001?
28. In what year did your facility achieve compliance with ISO 14001?
29. Do you think that government inspections are effective ways for inducing individual chemical facilities to comply with permitted water discharge limits?
    – Definitely Not
    – Probably Not
    – Probably Yes
    – Definitely Yes
30. Do you think it matters whether the inspections are performed by a state regulator or the EPA?
    – Definitely Not
    – Probably Not
    – Probably Yes
    – Definitely Yes
31. Which type of inspection is more effective?
    – State
    – Federal
32. Do you think that the imposition of monetary fines is an effective way for inducing individual chemical facilities to comply with permitted water discharge limits?
    – Definitely Not
    – Probably Not
    – Probably Yes
    – Definitely Yes
33. Assuming the fines are the same size, does it matter whether the monetary fines are state or federal?
    – Definitely Not
    – Probably Not
    – Probably Yes
    – Definitely Yes
34. Which type of monetary fine is more effective?
    – State
    – Federal

35. Regarding federal monetary fines, does it matter whether it is an EPA administrative fine or a federal civil court fine, assuming the fines are the same size?
    – Definitely Not
    – Probably Not
    – Probably Yes
    – Definitely Yes
36. Which type of federal monetary fine is more effective?
    – EPA Administrative Fine
    – Federal Civil Court Fine
37. Do you think that the imposition of a federal injunctive relief sanction is an effective way for inducing individual chemical facilities to comply with permitted water discharge limits in future situations?
    – Definitely Not
    – Probably Not
    – Probably Yes
    – Definitely Yes
38. Do you think that a federal requirement to complete a supplemental environmental project is an effective way for inducing individual chemical facilities to comply with permitted water discharge limits?
    – Definitely Not
    – Probably Not
    – Probably Yes
    – Definitely Yes

4A.2. PRESENCE OF LIMITS ACROSS VARIOUS SUBSAMPLES

This second section of the appendix to Chapter 4 assesses the presence of limits across various subsamples within the TSS sample and then within the BOD sample.

4A.2.1. Presence of TSS Limits Across Various Subsamples

This section assesses the presence of limits across various subsamples within the TSS sample.

First, it assesses the presence of limits across subsamples based on chemical manufacturing subsectors: organic chemical manufacturing, inorganic chemical manufacturing, and other chemical manufacturing. The presence of an operative TSS concentration limit differs across the sub-sectors; while 41 percent of the inorganic chemical subsector faces an operative limit, only 27 percent of the organic chemical subsector faces an operative limit. The presence of an operative quantity limit differs between the organic chemical subsector (89 percent) and the two remaining subsectors (61 percent for inorganic chemicals and 63 percent for "other" chemicals). Perhaps the best guide is the presence of any operative limit, which is clearly different in the organic chemical subsector. For this exceptional subsector, only 5 percent of facilities lack an operative limit in a given month. In contrast, 20 percent of the inorganic subsector lacks any operative limit. The organic chemical manufacturing subsector is clearly more likely to face only a quantity limit (68 percent) than are the other subsectors (39 percent for inorganic, 53 percent for "other"). The "other" chemical manufacturing subsector is less likely to face both a concentration and quantity limit (10 percent) than are the other subsectors (22 percent for inorganic, 21 percent for organic).

Next we assess the presence of limits across subsamples based on facility size. The presence of operative concentration limits does not differ due to facility size (32 percent versus 28 percent). However, the presence of quantity limits seems greater in larger facilities (88 percent versus 67 percent). The latter conclusion holds also for the presence of either a concentration or a quantity limit (94 percent versus 82 percent). Lastly, relative to smaller facilities, larger facilities are more likely to face only a TSS quantity limit (66 percent versus 50 percent). The other categories seem comparable between the two groups.

Lastly, this section assesses the presence of limits across subsamples based on time. The presence of an operative quantity limit grows from 74 percent to 77 percent, with a peak of 79 percent in 2001. The presence of an operative concentration limit grows from 26 percent to 33 percent monotonically. The presence of either limit grows from 83 percent to 89 percent, with a peak of 90 percent in 2001. The presence of both limits grows almost monotonically from 17 percent in 1999 to 20 percent in 2003.

4A.2.2. Presence of BOD Limits Across Various Subsamples

This section assesses the presence of limits across various subsamples within the BOD sample. First it assesses the presence of limits across subsamples based on chemical manufacturing subsectors. Within the full BOD sample, inorganic chemical manufacturing facilities represent 16 percent of the sample (561 observations), organic chemical facilities represent 65 percent (2,244 observations), and "other" chemical facilities represent

19 percent (663 observations). Clearly, organic chemical facilities are most likely to face some type of operative limit (95 percent); at the other extreme, inorganic chemical facilities are least likely to face some type of operative limit (77 percent); in between, 83 percent of "other" chemical facilities face an operative limit.

Next we assess the presence of limits across subsamples based on facility size. The presence of any operative limit does not appear to differ between smaller facilities (89 percent) and larger facilities (91 percent), with 1,734 observations in each subsample.

Finally we assess the presence of limits across subsamples based on time, with an exclusive focus on the presence of either an operative concentration or a quantity limit. The presence changes very little over time, rising from 88 percent in 1999 to a peak of 91.3 percent in 2001, then falling slightly to 90.7 percent by 2003.

Environmental Behavior:
Facilities' Efforts to Comply with
Discharge Limits

5.1. INTRODUCTION

Chapter 4 describes the imposition of discharge limits on polluting facilities. This chapter describes the environmental behavior facilities show as they expend effort to comply with the imposed discharge limits. With both descriptions in place, we probe the obvious connection between these two elements, which represents one of our research questions: how do imposed discharge limits affect environmental behavior? As our general hypothesis, we test whether tighter limits prompt better environmental behavior. Based on our testing, we interpret our answer within the following light: if environmental behavior improves as limits tighten, then limits would appear effective at prompting better behavior.

As described in the preceding chapter, our analysis focuses on total suspended solids (TSS) and biological oxygen demand (BOD) discharge limits and measurements. While some of the described behavioral elements relate directly to TSS or BOD, most of these elements relate generally to a variety of wastewater pollutants, if not a variety of pollutants emitted into other media (for example, air-pollutant emissions). We claim that each and every examined behavioral element has at least the potential to improve the management of TSS and BOD discharges, while acknowledging that the potential may substantially vary across the various behavioral elements.

5.2. BEHAVIORAL DATA GATHERED BY THE SURVEY

In this chapter, we analyze the behavioral elements gathered by the survey of chemical manufacturing facilities. In this section of the chapter, we describe these behavioral elements for the entire sample, without any references to other related factors, such as facility size. In Section 5A.1 of the appendix to this chapter, we depict the same data with reference to certain related factors: facility location, broad manufacturing sector, facility size, and time. In Section 5A.2 of the appendix, we assess variation in behavior across time for a given facility. In Section 4A.1 of the appendix to Chapter 4, we provide a full listing of the survey questions that provide information on environmental behavior, along with other information analyzed in this book.

Clearly, we explore a variety of environmental behavioral measures. First, we explore the quantity and quality of labor effort devoted to environmental management. We measure quantity based on two classifications: the overall number of environmental employees and the number of employees devoted to wastewater management. In both cases, we explore the absolute number of employees and the ratio of environmentally or wastewater-related employees to the overall number of employees working at a particular facility. Table 5.1 summarizes the absolute and relative number of environmental employees, which are measured on an annual basis for each year of the three-year period between 1999 and 2001. As shown, the median facility employs 60 environmental employees, which represent 2 percent of overall employees working for an individual facility. Table 5.1 also displays the absolute and relative number of wastewater employees, which are measured for the twelve-month period preceding survey completion.

We measure the quality of labor effort based on the employees' education and work experience, which are summarized in Table 5.1. As shown, the median facility employs environmental employees that are well educated: 50 percent of environmental employees are at least college educated. Moreover, the median facility hires environmental employees with much work experience; the normal environmental employee has twelve years of work experience. We also measure the quality of labor effort based on the provision of training, its quality, and its quantity. Table 5.1 displays the quality of training, in which external training is presumed better than internal training. Interestingly, only 1 percent of the facilities provide no training of any kind. In contrast, 42 percent of the facilities provide the highest quality of training. Table 5.1 also summarizes the quantity of training days provided by facilities and attendance at these provided training

TABLE 5.1

Depiction of environmental behavioral measures: Full sample

Table 5.1.a. Qualitative measures[a]

Behavior	Category	Frequency	Percentage
Training: Quality	No training	1	1.03
	Internal, in-house training	24	24.74
	External training	31	31.96
	Both in-house and external training	41	42.27
TSS Treatment Presence (N=95)	Absent	11	11.58
	Present	84	88.42
TSS Treatment Extent (N=95)	No treatment	11	11.58
	Coagulation plus sedimentation	27	28.42
	Sedimentation in effluent pond	8	8.42
	Sedimentation in final clarifier	36	37.89
	Granular media filtration	10	10.53
	Membrane filtration	3	3.16
BOD Treatment Presence (N=96)	Absent	18	18.75
	Present	78	81.25
BOD Treatment Extent (N=96)	No treatment	18	18.75
	Solids removal	13	13.54
	Stabilization pond treatment	8	8.33
	Trickling filter biological treatment	2	2.08
	Activated sludge biological treatment	51	53.13
	Chemical oxidation	2	2.08
	Carbon adsorption	2	2.08
Treatment Upgrade: Presence (N=94)	No	31	32.98
	Yes	63	67.02

Treatment Upgrade: Count of Years (N=94)	No upgrade	31	32.98
	Upgrade but no years indicated	3	3.19
	1 year	41	43.62
	2 years	8	8.51
	3 years	5	5.32
	4 years	6	6.38
Treatment Upgrade: Designed to Reduce Wastewater (N=92)	No	44	47.83
	Yes	48	52.17
Monitoring Program (N=96)	No monitoring (except final discharge)	9	9.38
	Treatment process monitoring	17	17.71
	Single-point sewer and treatment process	22	22.92
	Multiple-point sewer and treatment process	48	50.00
Audit: Team Composition (N=94)	No internal audits	1	1.09
	Visiting corporate and/or consultants only	10	10.87
	Teams of various facility employees	23	25.00
	Teams of facility employees and visiting corporate staff and/or consultants	58	63.04
Audit: Classification Protocol (N=94)	No audits	1	1.06
	Only noncompliance findings noted	20	21.28
	Two categories used: noncompliance versus other	26	27.66
	Findings placed into priority categories of vulnerability for noncompliance	23	24.47
	Findings ranked by degree of vulnerability for noncompliance	24	25.53
External Consultants: Use (N=96)	No	21	21.88
	Yes	75	78.13
Requested Assistance from Wastewater Regulator (N=95)	No	71	74.74
	Yes	24	25.26

(continued)

TABLE 5.1 *(continued)*

Table 5.1.a. Qualitative measures[a]

Behavior	Category	Frequency	Percentage
Compliance Assistance Meets Facility's Environmental Employees' Needs (N=89)	Never	15	16.85
	Some of time	20	22.47
	Most of time	24	26.97
	Always	30	33.71
Importance Management Places on Role of General Plant Employees in Identifying and Correcting Conditions Leading to Noncompliance (N=96)[b]	Somewhat important	6	6.25
	Important	10	10.42
	Very important	80	83.33
Management's Steps to Communicate Environmental Performance Goals and Targets to Facility Employees	0	2	2.06
	1	14	14.43
	2	11	11.34
	3	19	19.59
	4	27	27.84
	5	22	22.68
	6	2	2.06
Management's Steps to Communicate Environmental Performance Goals and Targets to External Parties	0	10	10.31
	1	22	22.68
	2	24	24.74
	3	17	17.53
	4	16	16.49
	5	8	8.25
ISO 14001 Compliance (N=90)	No	79	87.78
	Yes	11	12.22

[a]The sample size equals 97 unless otherwise noted in the most left-hand column.

[b]No facility chose the category of "Not important."

BOD, biological oxygen demand; TSS, total suspended solids.

Table 5.1.b. Quantitative measures

Behavior	N	Mean	25th Percentile	Median	75th Percentile
Environmental Employees: Absolute	280	101.139	14	60	131.5
Environmental Employees: Relative to Overall Employees	273	0.054	0.009	0.019	0.049
Wastewater Employees: Absolute	95	11.963	4	6	12
Wastewater Employees: Relative to Overall Employees	92	0.103	0.013	0.033	0.062
Environmental Employees' Education	94	54.660	20	50	95
Environmental Employees' Work Experience	95	13.474	10	12	15
Training Days	91	9.944	3	5	10
Training Attendance	94	89.968	100	100	100
Audit: Annual Count	262	5.718	1	1	4
External Consultants: Annual Expenditures—Absolute	88	33,540	1,000	10,000	30,000
External Consultants: Annual Expenditures—Relative to Overall Employees	87	280.85	5.77	50.00	200.00

sessions, as measured by the percentage of employees who generally attend. As shown, the median facility provides five days of training and facilitates 100 percent attendance.[1]

Second, in contrast to labor effort, we measure the provision of capital equipment. We measure the installation and use of treatment technologies and the extent of this treatment; we explore TSS removal and BOD removal separately. Consider initially TSS treatment. We contrast the presence of any TSS treatment technology with its absence. As shown in Table 5.1, 88 percent of the facilities employ at least some form of total suspended solids (TSS) treatment. We also distinguish across the various forms of TSS treatment, which are shown in order of increasing effectiveness, with membrane filtration as the most effective TSS treatment technology.[2] As for biological oxygen demand (BOD) treatment, 81 percent of facilities employ at least some type of BOD treatment technology. We also distinguish across the various forms of BOD treatment, which are shown in order of increasing effectiveness, with carbon adsorption as the most effective BOD treatment technology.[3] We also measure any improvements or upgrades to these treatment technologies during the three-year period prior to survey completion. We first assess the presence of any type of upgrade over this three-year period, as shown in Table 5.1. Then we assess the number of calendar years in which facilities performed these upgrades over this same three-year period. Since the three-year period may include four calendar years, the maximum count of calendar years is four rather than three. (Recall that facilities may have completed the survey in either 2002 or 2003. For facilities completing the survey in 2002, we consider the calendar years between 1999 and 2002, inclusively; for facilities completing the survey in 2003, we consider the calendar years between 2000 and 2003, inclusively.) As shown in Table 5.1, 44 percent of the facilities perform an upgrade in only a single year during the preceding three-year period. However, 12 percent of the facilities perform an upgrade in three or more calendar years. From this broad set of upgrades, we next identify the presence of an upgrade designed to reduce wastewater discharges. As shown in Table 5.1, while 67 percent of the facilities upgrade their treatment operations in general, only 52 percent upgrade their operations in order to reduce wastewater discharges.

Third, we assess facilities' efforts to understand their environmental management concerns. We assess the extent of a facility's monitoring program. As shown in Table 5.1, only 9 percent of the facilities fail to operate a monitoring program. In contrast, half of the facilities operate the most extensive monitoring program, involving multiple-point sewer and treatment process monitoring. More important, we assess facilities' effort to audit their operations. We measure both the quantity and quality of self-audits. We measure the quantity of self-audits on an annual basis over the three-year period between 1999 and 2001. As shown in Table 5.1, the median facility conducts one audit per year, while the average facility conducts almost six. However, 11 percent of the facilities conducted no audits during this three-year period. In contrast, 43 percent audited their facility annually, 13 percent audited quarterly, 10 percent audited monthly, and 5 percent audited weekly. Of course, the intensity of each audit may substantially differ between audits conducted annually and those conducted weekly. With this variation in mind, we also measure the quality of self-audits based on the audit team's composition and the extent of the audit classification protocol. For the former dimension of quality, we presume audit teams composed of only facility employees are better than teams composed of only visiting corporate employees or external consultants, or both, while teams composed of both elements are best.[4] As shown in Table 5.1, 63 percent of the facilities employ the highest-quality teams, while only 11 percent of the facilities employ the lowest-quality teams. The other audit quality measure displays more variation across the quality categories. For example, 21 percent of facilities employ the lowest-quality classification protocol, while 26 percent employ the highest-quality classification protocol.

Fourth, we assess facilities' efforts to seek insight from others regarding their environmental management efforts. Facilities may hire external consultants. Table 5.1 summarizes the use of external consultants in the three-year period preceding survey completion. As shown, 78 percent use external consultants. As a more comprehensive measure, we assess the annual amount of money spent on external consultants, in both an absolute sense and relative to a facility's size (as measured by the number of overall employees working at the facility).[5] In contrast to hired consultants, facilities may seek assistance and insight from their regulators. As shown in Table 5.1, 25 percent of the facilities at some point during the three-year period preceding survey completion contacted their wastewater regulator for assistance on how to reduce their wastewater discharges. These efforts to gather information and insight may be successful or not. We assess how frequently compliance-assistance materials, including those from consultants and regulators, meet the day-to-day needs of facility's environmental employees. As shown in Table 5.1, 34 percent of the facilities believe that their compliance materials always met their needs.

Fifth, we assess indicators of facilities' degree of concern over environmental management and to what extent these concerns translate into goals that are communicated to the facility's employees. We measure the level of importance management places on the role of general plant employees in identifying and correcting conditions that may lead to noncompliance. As shown in Table 5.1, 83 percent of the facilities claim that their management considers this role very important. In contrast, no facilities claim that their management believes that this role is not important.

We also assess the degree of effort expended by management to communicate environmental performance goals and targets to facility employees, measured by the number of different categories of steps taken by management. In total, we consider six categories: (1) regularly scheduled meetings or briefings; (2) regular memorandums, newsletters, or e-mails; (3) public display board; (4) general plant worker training or orientation session; (5) review item for new design or construction projects; (6) other. Only 2 facilities (2 percent) take no steps. At the other end, only 2 facilities take steps in all six categories. In between, 72 percent of the facilities take steps in at least three internal communication categories. While our assessment is able to explore the breadth of steps taken, it does not control for the depth or intensity of each step.

Sixth, in contrast to communication to facility employees, we also assess communication of environmental performance goals and targets to external parties, such as community leaders, again measured by the number of different categories of steps taken by management. In total, we consider six categories: (1) press releases, (2) advertising, (3) open houses, (4) focus groups, (5) Web sites or Internet messages, and (6) other. As shown in Table 5.1, 10 percent of the facilities take no steps, while no facilities take steps in all six categories. In between, 42 percent of the facilities take steps in at least three external communication categories. As with internal communication, our assessment of external communication explores only the breadth of steps taken rather than the depth of each step.

Seventh, we assess a broad indicator of a well-organized environmental management system: compliance with International Standards Organization (ISO) 14001 standards. As shown in Table 5.1, 12 percent of the facilities were compliant as of the time of the survey. Of course, facilities differ in their timing. Of the 11 facilities, 10 achieved ISO 14001 compliance during the sample period. Of these, 5 facilities (45 percent) achieved compliance in 2000, 3 facilities (27 percent) in 2001, and 2 facilities (18 percent) in 2002. One facility (9 percent) achieved compliance in 1996.

5.3. THE EFFECTS OF DISCHARGE LIMITS ON ENVIRONMENTAL BEHAVIOR

Given this depiction of the surveyed elements of environmental behavior, we next attempt to answer one of our research questions: how do imposed discharge limits affect environmental behavior? As our general hypothesis, we test whether tighter limits prompt better environmental behavior, which implies that limits are effective at prompting better behavior.

In order to test this general hypothesis, we examine the effects of discharge limits on the various measures of environmental behavior. For this examination, we discern between qualitative measures of behavior, which are identified by categories (for example, presence or absence of ISO 14001 compliance), and quantitative measures of behavior, which are identified by a number (for example, number of environmental employees). The former set of measures includes both qualitative measures that include only two categories and qualitative measures that include more than two categories. In order to interpret our analysis with more clarity, we reformulate each qualitative measure of multiple categories into a broader qualitative measure of only two categories.

To elaborate, we simplify our analysis of environmental management practices measured in ordered categories. For example, the quality of training provided to facility employees includes the following categories: no training, internal training only, external training only, and both internal and external training. For each relevant management practice, we classify the multiple categories into two broader categories. Based on the relevant data, we find it helpful to distinguish the most exemplary category (for example, both internal and external training) from the less than exemplary categories (for example, no training, internal training only, external training only). This distinction proves meaningful for interpretation and reasonably divides the prevalence of exhibited management practices (that is, a reasonable portion of facilities lie within each broader category). The two exceptions apply to BOD treatment technology and TSS treatment technology. For both treatment technologies, we first distinguish between the presence of any treatment technology and its absence and second distinguish between greater treatment technologies and less treatment technologies based on their removal effectiveness. In the case of removal effectiveness, for BOD treatment, we classify the following technologies as greater: activated sludge biological treatment, chemical oxidation, and carbon absorption; we classify the following technologies as lesser: solids removal, stabilization pond treatment, and trickling filter biological treatment; clearly, we classify the absence of treatment as a lesser technology. For TSS treatment, we classify the following technologies as greater: sedimentation in final clarifier, granular media filtration, and membrane filtration; while classifying the following technologies as lesser: coaguluation plus sedimentation and sedimentation in effluent pond; clearly, we classify the absence of treatment as a lesser technology.

In order to test our hypothesis that tighter limits prompt better environmental behavior, we assess whether the extent of environmental behavior differs between two subsamples: facilities facing below-median discharge limits and facilities facing above-median discharge limits. For quantitative measures of environmental behavior, we also assess the correlation between the level of a discharge limit and the extent of environmental behavior.

Readers less interested in the details of our statistical analysis than in the results of our statistical analysis may wish to skip over the following five paragraphs or even to the beginning of Section 5.4, in which we interpret the results of our hypothesis testing.

For all of the qualitative measures—originally with only two categories and reformulated to include only two categories—we assess whether the distribution of the behavioral measure differs between two subsamples: facilities facing below-median discharge limits and facilities facing above-median discharge limits. We first identify the distributions in each subsample and then compare the distributions. Finally, we employ a two-sample test in order to assess whether any identified difference in the subsample-specific distributions is statistically significant.[6]

For the quantitative measures of environmental behavior, we also assess whether the distribution of the behavioral measure differs between the subsample of facilities facing below-median discharge limits and the subsample of facilities facing above-median discharge limits. Similar to the qualitative measures of behavior, we identify the distributions in each subsample, compare the distributions, and then employ two-sample tests— the Wilcoxon two-sample test and a median two-sample test—in order to assess whether any identified difference in the subsample-specific distributions is statistically significant. We report only the former test results because the latter test results generate identical conclusions with one noted exception. (Section 5A.3 of the appendix describes an alternative two-sample test available for quantitative measures.)

Two-sample tests compare the distributions associated with two subsamples. In particular, these tests assess whether the distribution of one subsample is located at higher levels of the variable in question than the distribution of the other subsample. The two subsamples are identified by a single factor. In this case, we identify the two subsamples based on the level of the discharge limit—whether it lies below or above the sample median discharge limit. Any significant difference between the two distributions based on the two subsamples provides evidence that the discerning factor, in this case the level of discharge limit, influences the distribution of the variable of interest, in this case the extent of environmental behavior. Specific to our hypothesis, if the distribution of environmental behavior associated with a lower limit (below-median-limit level) lies significantly above the distribution of environmental behavior associated with a higher limit (above-median-limit level), then tighter limits would appear to prompt better environmental behavior. We deem a difference between two distributions statistically significant when our analysis is able to reject the null hypothesis of zero difference at a significance level of 10 percent or less, indicating that the likelihood of incorrectly rejecting the null hypothesis lies at or below 10 percent. This situation is revealed by a p-value of 0.10 or less. We deem a difference between two distributions statistically insignificant when our analysis is not able to reject the null hypothesis of zero difference at a significance level of 10 percent or less, indicating that the likelihood of incorrectly rejecting the null hypothesis lies above 10 percent. This situation is revealed by a p-value greater than 0.10.

In addition to the two-sample tests, for the quantitative behavioral measures, we also assess the correlation between a behavioral measure and a discharge-limit level by calculating the Pearson correlation coefficients associated with the relevant pairwise comparisons of individual behavioral measures and discharge limits. Each correlation captures whether a discharge-limit level and behavioral measure vary together in a systematic fashion. For our exploration, we first assess the signs of these correlations in order to explore the types of relationships between each behavioral measure and discharge limit. A positive correlation indicates that the extent of behavior and level of the discharge limit generally vary together in the same direction; that is, when one is high, the other is high, and vice versa. A negative correlation indicates that the extent of behavior and level of the discharge limit generally vary together in the opposite direction; that is, when one is high, the other is low, and vice versa. We also assess the magnitude of these correlations in order to explore the strength of the relationships between environmental behavior and discharge limits. Correlation magnitudes above 0.50 generally indicate a reasonably strong relationship between the pairing of environmental behavior and discharge limit. In addition, we assess the statistical significance of these correlations, as captured by the p-values associated with the correlations, in order to test whether the calculated magnitudes are distinguishable from zero; a zero magnitude implies no relationship between the individual environmental behavioral measure and discharge-limit level. The calculated level of statistical significance, reflected in the associated p-value, stems from a comparison of the correlation magnitude and the underlying variance of the two factors, akin to a signal-to-noise ratio. We deem a correlation statistically significant when our analysis is able to reject the null hypothesis of a zero correlation at a significance level of 10 percent or less, indicating that the likelihood of incorrectly rejecting the null hypothesis lies at or below 10 percent. This situation is revealed by a p-value of 0.10 or less. We deem a correlation statistically insignificant when our analysis is not able to reject the null hypothesis of a zero correlation at a significance level of 10 percent or less, indicating that the likelihood of incorrectly rejecting the null hypothesis lies above 10 percent. This situation is revealed by a p-value greater than 0.10.

As neither fish nor fowl, we modify our analysis of the count of different internal communication methods employed by a facility and the count of different external communications methods employed by a facility. As one modification, for these two management practices, we use the midpoint of the distribution (three in both cases) to divide facilities into two broad categories: fewer methods and more methods. As another modification, we simply treat the communication-method counts as quantitative measures of the extent to which facilities attempt to communicate internally or externally.

In the preceding chapter, we describe and interpret four forms of discharge limit: concentration, concentration-equivalent, quantity, and production-adjusted quantity. As argued in Chapter 4, we believe that the concentration-equivalent limit and the production-adjusted quantity limit are the most informative limits. Consequently, we examine only the effects of these two limits on environmental behavior.

Lastly, for our examination of discharge limits' effects on environmental behavior, we must aggregate our data on discharge limits to a frequency that is identical to our data on behavior. Recall that some behavioral forms are measured for each of three calendar years between 1999 and 2001, inclusively. Other behavioral forms are measured for the twelve-month period preceding the completion of the survey. The remaining behavioral forms are measured for the three-year period preceding the completion of the survey. Since the time frame differs across three broad categories of behavioral forms, our exploration of discharge limits involves three different time frames. Specifically, we assess whether the distribution of the behavioral measure differs between two subsamples: facilities facing below-median discharge limits and facilities facing above-median discharge limits. This division is specific to a particular time frame.[7]

For each type of pollutant—TSS and BOD—and for each type of limit—concentration-equivalent and production-adjusted quantity—we summarize the monthly data on discharge limits using the minimum-discharge-limit level for a particular facility over the relevant period: calendar year, twelve months preceding survey, or three years preceding survey. Use of the facility-specific and time-period-specific discharge-limit minimum is consistent with our focus on the month-specific discharge-limit minimum when multiple limit levels apply to a single facility for a given pollutant in a particular measurement form (concentration versus quantity). Section 5A.5 of the appendix discusses an alternative approach: summarize the monthly data on discharge limits using the median-discharge-limit level for a particular facility over the relevant period.

Given all of this framing, we explore the effect of discharge limits on various forms of environmental behavior. We assess first TSS limits, since this pollutant is more prominent in the sample, while starting with production-adjusted quantity limits, because quantity limits are more prevalent in the sample than concentration limits. After examining both TSS production-adjusted quantity limits and concentration-equivalent limits, we attempt to draw a conclusion for the influence of TSS limits in general on each form of environmental behavior. Then we assess BOD limits, again starting with production-adjusted quantity limits. After examining both BOD production-adjusted quantity limits and concentration-equivalent limits, we attempt to draw a conclusion for the influence of BOD limits in general on each form of environmental behavior. Finally, we attempt to draw a conclusion for the influence of discharge limits in general on each form of environmental behavior.

In all cases, we attempt to identify evidence that either supports or rejects our overall hypothesis that tighter limits induce facilities to employ stronger environmental behavior.

5.4.1. Hypothesis Testing: TSS Limits

5.4.1.1. Hypothesis Testing: TSS Quantity Limits. First, we consider the effect of TSS production-adjusted quantity limits on the number of environmental employees, measured in both an absolute sense and relative to the overall number of employees working at an individual facility. As shown in Table 5.2.a, facilities facing tighter limits (below-median-limit levels) hire fewer environmental employees, absolutely and relative to overall employees, based on the mean values (and median values), which is inconsistent with our hypothesis. However, the difference is quite small when scaling the number of environmental employees by the number of overall employees. In both cases, the difference proves statistically insignificant. In contrast to the two-sample assessment, correlations indicate that as limit levels fall (limits tighten), facilities employ more environmental employees, absolutely and relative to the number of overall employees ($\rho = -0.051$, $\rho = -0.016$), consistent with our hypothesis. Nevertheless, both relationships prove statistically insignificant ($p = 0.484$, $p = 0.830$). Thus, regardless of the method for exploring the effect of

TSS production-adjusted quantity limits on the number of environmental employees and the manner of measuring environmental employees (absolute versus relative), TSS quantity limits do not appear to influence the number of environmental employees, thus providing no support for our hypothesis that tighter limits induce facilities to employ more environmental employees.

As an alternative measure of quantity of labor effort devoted to environmental management, we explore the effect of TSS production-adjusted quantity limits on the number of employees devoted to wastewater management, as measured by the absolute number of employees and the ratio of wastewater-related employees to overall employees working at the particular facility. As shown in Table 5.2.a, facilities facing tighter limits hire more wastewater employees, in an absolute sense, based on mean values (and medians), consistent with our hypothesis. However, facilities facing tighter limits hire fewer wastewater employees, relative to overall employees, based on mean values (and medians), inconsistent with our hypothesis. In both cases, the difference proves statistically insignificant. Correlations indicate that as limit levels fall (limits tighten), facilities employ more wastewater employees, absolutely and relative to the number of overall employees ($\rho = -0.089$, $\rho = -0.060$), consistent with our hypothesis. Nevertheless, both relationships prove statistically insignificant ($p = 0.480$, $p = 0.634$). Thus, regardless of the method for exploring the effect of TSS production-adjusted quantity limits on the number of wastewater employees and the manner of measuring wastewater employees (absolute versus relative), TSS quantity limits do not appear to influence the number of wastewater employees, thus providing no support for our hypothesis that tighter limits induce facilities to employ more wastewater employees.

We next explore the effect of TSS production-adjusted quantity limits on the quality of labor effort, as measured by the employees' education and experience. As shown in Table 5.2.a, facilities facing tighter limits employ a greater proportion of employees with at least a college degree, based on mean values (and medians), consistent with our hypothesis. However, the difference in means is very small (55.9 versus 55.5). In the case of work experience, facilities facing tighter limits employ employees with more work experience, based on the mean values, consistent with our hypothesis. However, the difference in means is very small: 13.48 versus 13.45. (The median value is identical in the two subsamples.) In both dimensions of labor quality, the difference between facilities facing tighter limits and those facing looser limits proves statistically insignificant. Correlations indicate that facilities hire a greater proportion of college-educated environmental employees as limits loosen ($\rho = 0.102$), inconsistent with our hypothesis, while facilities hire workers with greater work

TABLE 5.2

Hypothesis testing: Effects of discharge limits on environmental behavior

Table 5.2.a. TSS production-adjusted quantity limits

Table 5.2.a.i. Qualitative measures of behavior

| Behavioral Measure | Category | Frequency (Subsample Conditional Percentage)[a] | | Test Statistic (p-value)[b] |
		Below-Median-Limit Subsample	Above-Median-Limit Subsample	
Training: Quality	Lower: none, internal only, external only	16	24	0.1212
	Higher: internal and external	17	9	(0.287)
		(51.52)	(27.27)	
TSS Treatment Presence	Absent	NS[c]	NS[c]	0.0005
	Present	NS[c]	NS[c]	(1.000)
TSS Treatment Extent	Lesser: none, coagulation + sedimentation, sedimentation in effluent pond	11	13	0.0251
	Greater: sedimentation in final clarifier, granular media filtration, membrane filtration	21	20	(1.000)
		(65.63)	(60.61)	
BOD Treatment Presence	Absent	NS[c]	NS[c]	0.0455
	Present	NS[c]	NS[c]	(0.999)
BOD Treatment Extent	Lesser: none, solids removal, stabilization pond, trickling filter	10	12	0.0303
	Greater: activated sludge, chemical oxidation, carbon adsorption	23	21	(1.000)
		(69.70)	(63.64)	

(continued)

TABLE 5.2 (*continued*)

Table 5.2.a. TSS production-adjusted quantity limits

Table 5.2.a.i. Qualitative measures of behavior

Behavioral Measure	Category	Frequency (Subsample Conditional Percentage)[a]		Test Statistic (p-value)[b]
		Below-Median-Limit Subsample	Above-Median-Limit Subsample	
Treatment Upgrade: Presence	No	10	14	0.0588
	Yes	24	20	(0.973)
		(70.59)	(58.82)	
Treatment Upgrade: Designed to Reduce Wastewater	No	15	19	0.0450
	Yes	17	15	(0.999)
		(53.13)	(44.12)	
Monitoring Program	Lesser: none (except final discharge), treatment process, single-point sewer and treatment process	13	14	0.0152 (1.000)
	Greater: multiple-point sewer and treatment process	20	19	
		(60.61)	(57.58)	
Audit: Team Composition	Lesser: none, visiting corporate/consultants, teams of facility employees	8	14	0.1008 (0.544)
	Greater: teams of visiting corporate/consultants and facility employees	24	17	
		(75.00)	(54.84)	
Audit: Classification Protocol	Lesser: none, noncompliance findings, noncompliance versus other, priority categories of vulnerability	26	25	0.0156 (1.000)
	Greater: ranked serially by vulnerability for noncompliance	6	7	
		(18.75)	(21.88)	

External Consultants: Use	No	9	7	0.0294
	Yes	25	27	(1.000)
		(73.53)	(79.41)	
Requested Assistance from Wastewater Regulator	No	26	27	0.0376
	Yes	9	6	(1.000)
		(25.71)	(18.18)	
Compliance Assistance Meets Facility's Environmental Employees' Needs	Lesser: never, some of time, most of time	23	19	0.0516
	Greater: always	10	13	(0.995)
		(30.30)	(40.63)	
Importance Management Places on Role of General Plant Employees in Identifying and Correcting Conditions Leading to Noncompliance	Lesser: somewhat important, important	NS[c]	NS[c]	0.1364
	Greater: very important	NS[c]	NS[c]	(0.172)
Management's Steps to Communicate Environmental Performance Goals and Targets to Facility Employees	Below midpoint	17	20	0.0455
	At or above midpoint	16	13	(0.999)
		(48.48)	(39.39)	
Management's Steps to Communicate Environmental Performance Goals and Targets to External Parties	Below midpoint	24	25	0.0152
	At or above midpoint	9	8	(1.000)
		(27.27)	(24.24)	
ISO 14001 Compliance	No	NS[c]	NS[c]	0.0323
	Yes	NS[c]	NS[c]	(1.000)

[a]The subsample conditional percentage is reported only for the category that represents either the presence of a behavioral measure or the greater extent of a behavioral measure. For example, in the subsample of facilities facing a below-median limit, the reported percentage indicates the number of observations where the facility employs the identified behavior or the greater extent of behavior, while also facing a below-median limit, relative to the number of all facilities facing a below-median limit.

[b]Test statistic applies to the comparison between the below-median limit subsample and above-median limit subsample.

[c]NS indicates that at least one cell in the cross-tabulation includes 3 or fewer observations if the cell is populated.

BOD, biological oxygen demand; TSS, total suspended solids.

(continued)

TABLE 5.2 (*continued*)

Table 5.2.a.ii. Quantitative measures of behavior

Behavioral Measure	Limit Subsample	N	Mean	Sign of Difference in Means[a]	Test Statistic (p-value)[b]
Environmental Employees: Absolute	Below median limit	94	88.649	+	−0.816
	Above median limit	97	133.620		(0.414)
Environmental Employees: Relative to Overall Employees	Below median limit	93	0.033	+	−0.795
	Above median limit	94	0.035		(0.427)
Wastewater Employees: Absolute	Below median limit	33	12.833	−	1.421
	Above median limit	33	8.167		(0.155)
Wastewater Employees: Relative to Overall Employees	Below median limit	32	0.086	+	−0.053
	Above median limit	33	0.100		(0.958)
Environmental Employees' Education	Below median limit	31	55.935	−	−0.102
	Above median limit	33	55.515		(0.919)
Environmental Employees' Work Experience	Below median limit	33	13.485	−	−0.096
	Above median limit	31	13.452		(0.924)
Training Days	Below median limit	31	10.210	+	−0.604
	Above median limit	33	10.589		(0.546)

Training Attendance	Below median limit	32	85.125	+	−0.849
	Above median limit	33	94.091		(0.396)
Audit: Annual Count	Below median limit	95	6.074	−	−0.590
	Above median limit	87	4.793		(0.555)
External Consultants: Annual Expenditures—Absolute	Below median limit	32	50,875	−	−0.095
	Above median limit	32	23,641		(0.924)
External Consultants: Annual Expenditures—Relative to Overall Employees	Below median limit	32	188.79	+	−0.920
	Above median limit	32	283.28		(0.357)
Management's Steps to Communicate Environmental Performance Goals and Targets to Facility Employees (count)	Below median limit	33	3.212	−	0.170
	Above median limit	33	3.152		(0.865)
Management's Steps to Communicate Environmental Performance Goals and Targets to External Parties (count)	Below median limit	33	2.364	−	−0.170
	Above median limit	33	2.364		(0.865)

[a]The difference in mean values = (mean of below median limit subsample) − (mean of above median limit subsample).

[b]Test statistic applies to the comparison between the below-median-limit subsample and above-median-limit subsample; it reports the Wilcoxon two-sample test statistic in the form of a normal approximation.

(continued)

Table 5.2.b. TSS concentration-equivalent limits

Table 5.2.b.i. Qualitative measures of behavior

TABLE 5.2 (continued)

Behavioral Measure	Category	Frequency (Subsample Conditional Percent)[a]		Test Statistic (p-value)[b]
		Below-Median-Limit Subsample	Above-Median-Limit Subsample	
Training: Quality	Lower: none, internal only, external only	15	5	0.0907 (0.906)
	Higher: internal and external	11 (42.31)	8 (61.54)	
TSS Treatment Presence	Absent	NS^c	NS^c	0.0569 (0.999)
	Present	NS^c	NS^c	
TSS Treatment Extent	Lesser: none, coagulation+sedimentation, sedimentation in effluent pond	15	5	0.1022 (0.822)
	Greater: sedimentation in final clarifier, granular media filtration, membrane filtration	10 (40.00)	8 (61.54)	
BOD Treatment Presence	Absent	NS^c	NS^c	0.0544 (0.999)
	Present	NS^c	NS^c	
BOD Treatment Extent	Lesser: none, solids removal, stabilization pond, trickling filter	13	4	0.0907 (0.906)
	Greater: activated sludge, chemical oxidation, carbon adsorption	13 (50.00)	9 (69.23)	
Treatment Upgrade: Presence	No	8	5	0.0307 (1.000)
	Yes	17 (68.00)	8 (61.54)	
Treatment Upgrade: Designed to Reduce Wastewater	No	13	8	0.0949 (0.893)
	Yes	14 (58.33)	5 (38.46)	

Variable	Category			
Monitoring Program	Lesser: none (except final discharge), treatment process, single-point sewer and treatment process	11	5	0.0181 (1.000)
	Greater: multiple-point sewer and treatment process	15 (57.69)	8 (61.54)	0.0689 (0.995)
Audit: Team Composition	Lesser: none, visiting corporate/consultants, teams of facility employees	NS[c]	NS[c]	
	Greater: teams of visiting corporate/consultants and facility employees	NS[c]	NS[c]	0.0477 (1.000)
Audit: Classification Protocol	Lesser: none, noncompliance findings, noncompliance versus other, priority categories of vulnerability	20	8	
	Greater: ranked serially by vulnerability for noncompliance	6 (23.08)	4 (33.33)	0.0214 (1.000)
External Consultants: Use	No	NS[c]	NS[c]	
	Yes	16	13	0.1908 (0.109)
Requested Assistance from Wastewater Regulator	No	NS[c]	NS[c]	
	Yes	11 (40.74)	0 (0.00)	0.0643 (0.998)
Compliance Assistance Meets Facility's Environmental Employees' Needs	Lesser: never, some of time, most of time	18	8	
	Greater: always	6 (25.00)	5 (38.46)	0.0181 (1.000)
Importance Management Places on Role of General Plant Employees in Identifying and Correcting Conditions leading to Noncompliance	Lesser: somewhat important, important	NS[c]	NS[c]	
	Greater: very important	NS[c]	NS[c]	

(continued)

TABLE 5.2 (*continued*)

Table 5.2.b. TSS concentration-equivalent limits

Table 5.2.b.i. Qualitative measures of behavior

| Behavioral Measure | Category | Frequency (Subsample Conditional Percent)[a] | | Test Statistic (p-value)[b] |
		Below-Median-Limit Subsample	Above-Median-Limit Subsample	
Management's Steps to Communicate Environmental Performance Goals and Targets to Facility Employees	Below midpoint At or above midpoint	8 18 (69.23)	6 7 (53.85)	0.0725 (0.987)
Management's Steps to Communicate Environmental Performance Goals and Targets to External Parties	Below midpoint At or above midpoint	22 4 (15.38)	8 5 (38.46)	0.1088 (0.745)
ISO 14001 Compliance	No Yes	NS[c] NS[c]	NS[c] NS[c]	0.0107 (1.000)

[a]The subsample conditional percentage is reported only for the category that represents either the presence of a behavioral measure or the greater extent of a behavioral measure. For example, in the subsample of facilities facing a below-median limit, the reported percentage indicates the number of observations where the facility employs the identified behavior or the greater extent of behavior, while also facing a below-median limit, relative to the number of all facilities facing a below-median limit.

[b]Test statistic applies to the comparison between the below-median-limit subsample and above-median-limit subsample.

[c]NS indicates that at least one cell in the cross-tabulation includes 3 or fewer observations if the cell is populated.

BOD, biological oxygen demand; TSS, total suspended solids.

Table 5.2.b.ii. Quantitative measures of behavior

Behavioral Measure	Limit Subsample	N	Mean	Sign of Difference in Means[a]	Test Statistic (p-value)[b]
Environmental Employees: Absolute	Below median limit	68	61.316	+	3.518
	Above median limit	29	138.36		(0.000)
Environmental Employees: Relative to Overall Employees	Below median limit	68	0.026	+	4.039
	Above median limit	29	0.069		(0.000)
Wastewater Employees: Absolute	Below median limit	26	13.962	–	0.135
	Above median limit	13	8.962		(0.893)
Wastewater Employees: Relative to Overall Employees	Below median limit	26	0.113	+	0.045
	Above median limit	13	0.153		(0.964)
Environmental Employees' Education	Below median limit	25	57.100	+	0.405
	Above median limit	13	64.308		(0.685)
Environmental Employees' Work Experience	Below median limit	25	12.360	–	–0.319
	Above median limit	13	11.000		(0.749)
Training Days	Below median limit	23	8.804	+	0.170
	Above median limit	13	13.692		(0.865)
Training Attendance	Below median limit	24	90.375	+	0.729
	Above median limit	13	97.308		(0.466)
Audit: Annual Count	Below median limit	60	10.967	–	–2.851
	Above median limit	28	5.250		(0.004)

(continued)

TABLE 5.2 (continued)

Table 5.2.b.ii. Quantitative measures of behavior

Behavioral Measure	Limit Subsample	N	Mean	Sign of Difference in Means[a]	Test Statistic (p-value)[b]
External Consultants: Annual Expenditures—Absolute	Below median limit	24	22,813	+	0.273
	Above median limit	13	26,462		(0.785)
External Consultants: Annual Expenditures—Relative to Overall Employees	Below median limit	24	173.85	+	-0.416
	Above median limit	13	362.35		(0.678)
Management's Steps to Communicate Environmental Performance Goals and Targets to Facility Employees (count)	Below median limit	26	3.885	-	-0.752
	Above median limit	13	3.462		(0.452)
Management's Steps to Communicate Environmental Performance Goals and Targets to External Parties (count)	Below median limit	26	2.192	+	1.072
	Above median limit	13	2.692		(0.284)

[a]The difference in mean values = (mean of above median limit subsample) − (mean of below median limit subsample).
[b]Test statistic applies to the comparison between the below-median-limit subsample and above-median-limit-subsample; it reports the Wilcoxon two-sample test statistic in the form of a normal approximation.

Table 5.2.c. BOD production-adjusted quantity limits
Table 5.2.c.i. Qualitative measures of environmental behavior

Behavioral Measure	Category	Frequency (Subsample Conditional Percent)[a]		Test Statistic (p-value)[b]
		Below-Median-Limit Subsample	Above-Median-Limit Subsample	
Training: Quality	Lower: none, internal only, external only	17	21	0.0690
	Higher: internal and external	12	8	(0.946)
		(41.38)	(27.59)	
TSS Treatment Presence	Absent	NS[c]	NS[c]	0.0517
	Present	NS[c]	NS[c]	(0.998)
TSS Treatment Extent	Lesser: none, coagulation + sedimentation, sedimentation in effluent pond	11	10	0.0111
	Greater: sedimentation in final clarifier, granular media filtration, membrane filtration	18	18	(1.000)
		(62.07)	(64.29)	
BOD Treatment Presence	Absent	NS[c]	NS[c]	0.0172
	Present	NS[c]	NS[c]	(1.000)
BOD Treatment Extent	Lesser: none, solids removal, stabilization pond, trickling filter	5	9	0.0690
	Greater: activated sludge, chemical oxidation, carbon adsorption	24	20	(0.946)
		(82.76)	(68.97)	
Treatment Upgrade: Presence	No	14	10	0.0521
	Yes	18	20	(0.996)
		(56.25)	(66.67)	

(continued)

TABLE 5-2 (continued)

Table 5.2.c. BOD production-adjusted quantity limits

Table 5.2.c.i. Qualitative measures of environmental behavior

Behavioral Measure	Category	Frequency (Subsample Conditional Percent)[a]		Test Statistic (p-value)[b]
		Below-Median-Limit Subsample	Above-Median-Limit Subsample	
Treatment Upgrade: Designed to Reduce Wastewater	No	18	13	0.0661
	Yes	13	16	(0.956)
		(41.94)	(55.17)	
Monitoring Program	Lesser: none (except final discharge), treatment process, single-point sewer and treatment process	12	13	0.0172
				(1.000)
	Greater: multiple-point sewer and treatment process	17	16	
		(58.62)	(55.17)	
Audit: Team Composition	Lesser: none, visiting corporate/consultants, teams of facility employees	8	12	0.0794
				(0.879)
	Greater: teams of visiting corporate/consultants and facility employees	20	15	
		(71.43)	(55.56)	
Audit: Classification Protocol	Lesser: none, noncompliance findings, noncompliance versus other, priority categories of vulnerability	23	21	0.0357
				(1.000)
	Greater: ranked serially by vulnerability for noncompliance	5	7	
		(17.86)	(25.00)	
External Consultants: Use	No	9	6	0.0484
	Yes	22	25	(0.999)
		(70.97)	(80.65)	

		Below-median-limit subsample	Above-median-limit subsample	Test statistic[b]
Requested Assistance from Wastewater Regulator	No	25	25	0.0000
	Yes	6	6	(1.000)
		(19.35)	(19.35)	
Compliance Assistance Meets Facility's Environmental Employees' Needs	Lesser: never, some of time, most of time	22	16	0.0908
	Greater: always	8	13	(0.716)
		(26.67)	(44.83)	
Importance Management Places on Role of General Plant Employees in Identifying and Correcting Conditions leading to NonCompliance	Lesser: somewhat important, important	NS[c]	NS[c]	0.0862
	Greater: very important	NS[c]	NS[c]	(0.782)
Management's Steps to Communicate Environmental Performance Goals and Targets to Facility Employees	Below midpoint	17	16	0.0172
	At or above midpoint	12	13	(1.000)
		(41.38)	(44.83)	
Management's Steps to Communicate Environmental Performance Goals and Targets to External Parties	Below midpoint	22	20	0.0345
	At or above midpoint	7	9	(1.000)
		(24.14)	(31.03)	
ISO 14001 Compliance	No	21	25	0.0110
	Yes	4	4	(1.000)
		(16.00)	(13.79)	

[a]The subsample conditional percentage is reported only for the category that represents either the presence of a behavioral measure or the greater extent of a behavioral measure. For example, in the subsample of facilities facing a below-median limit, the reported percentage indicates the number of observations where the facility employs the identified behavior or the greater extent of behavior, while also facing a below-median limit, relative to the number of all facilities facing a below-median limit.

[b]Test statistic applies to the comparison between the below-median-limit subsample and above-median-limit subsample.

[c]NS indicates that at least one cell in the cross-tabulation includes 3 or fewer observations if the cell is populated.

BOD, biological oxygen demand; TSS, total suspended solids.

(continued)

TABLE 5.2 (continued)

Table 5.2.c.ii. Quantitative measures of environmental behavior

Behavioral Measure	Limit Subsample	N	Mean	Sign of Difference in Means[a]	Test Statistic (p-value)[b]
Environmental Employees: Absolute	Below median limit	86	125.47	–	0.176
	Above median limit	87	113.41		(0.860)
Environmental Employees: Relative to Overall Employees	Below median limit	82	0.033	+	-1.195
	Above median limit	87	0.040		(0.232)
Wastewater Employees: Absolute	Below median limit	29	9.172	+	-0.070
	Above median limit	29	9.793		(0.944)
Wastewater Employees: Relative to Overall Employees	Below median limit	28	0.107	–	-1.174
	Above median limit	29	0.075		(0.241)
Environmental Employees' Education	Below median limit	29	54.448	+	0.410
	Above median limit	28	57.643		(0.682)
Environmental Employees' Work Experience	Below median limit	29	13.345	–	-0.392
	Above median limit	27	13.111		(0.695)
Training Days	Below median limit	28	9.821	+	1.351
	Above median limit	27	11.565		(0.177)
Training Attendance	Below median limit	29	83.241	+	1.064
	Above median limit	28	93.750		(0.287)
Audit: Annual Count	Below median limit	86	7.895	–	-0.636
	Above median limit	81	4.864		(0.525)

External Consultants: Annual Expenditures—Absolute	Below median limit	30	28,333	+	0.787
	Above median limit	28	37,125		(0.431)
External Consultants: Annual Expenditures—Relative to Overall Employees	Below median limit	30	118.21	+	1.224
	Above median limit	28	259.33		(0.221)
Management's Steps to Communicate Environmental Performance Goals and Targets to Facility Employees (count)	Below median limit	29	3.069	+	−0.143
	Above median limit	29	3.103		(0.887)
Management's Steps to Communicate Environmental Performance Goals and Targets to External Parties (count)	Below median limit	29	2.241	+	−0.887
	Above median limit	29	2.552		(0.375)

[a]The difference in mean values = (mean of above median limit subsample) − (mean of below median limit subsample).

[b]Test statistic applies to the comparison between the below median limit subsample and above median limit subsample; it reports the Wilcoxon Two-Sample Test statistic in the form of a normal approximation.

(continued)

TABLE 5.2 *(continued)*

Table 5.2.d. BOD concentration-equivalent limits

Table 5.2.d.i. Qualitative measures of environmental behavior

Behavioral Measure	Category	Frequency (Subsample Conditional Percent)[a]		Test Statistic (p-value)[b]
		Below-Median-Limit Subsample	Above-Median-Limit Subsample	
Training: Quality	Lower: none, internal only, external only	8	8	0.0220
	Higher: internal and external	6	5	(1.000)
		(42.86)	(38.46)	
TSS Treatment Presence	Absent	NS[c]	NS[c]	0.0385
	Present	NS[c]	NS[c]	(1.000)
TSS Treatment Extent	Lesser: none, coagulation+sedimentation, sedimentation in effluent pond	5	5	0.0000 (1.000)
	Greater: sedimentation in final clarifier, granular media filtration, membrane filtration	8	8	
		(61.54)	(61.54)	
BOD Treatment Presence	Absent	NS[c]	NS[c]	0.0329
	Present	NS[c]	NS[c]	(1.000)
BOD Treatment Extent	Lesser: none, solids removal, stabilization pond, trickling filter	NS[c]	NS[c]	0.0659 (0.999)
	Greater: activated sludge, chemical oxidation, carbon adsorption	NS[c]	NS[c]	
Treatment Upgrade: Presence	No	4	6	0.0879
	Yes	10	7	(0.985)
		(71.43)	(53.85)	

Treatment Upgrade: Designed to Reduce Wastewater	No	7	8	0.0385
	Yes	6 (46.15)	5 (38.46)	(1.000)
Monitoring Program	Lesser: none (except final discharge), treatment process, single-point sewer and treatment process	6	7	0.0549
	Greater: multiple-point sewer and treatment process	8 (57.14)	6 (46.15)	(1.000)
Audit: Team Composition	Lesser: none, visiting corporate/consultants, teams of facility employees	NS[c]	NS[c]	0.1729
	Greater: teams of visiting corporate/consultants and facility employees	NS[c]	NS[c]	(0.443)
Audit: Classification Protocol	Lesser: none, noncompliance findings, noncompliance versus other, priority categories of vulnerability	NS[c]	NS[c]	0.0178
	Greater: ranked serially by vulnerability for noncompliance	NS[c]	NS[c]	(1.000)
External Consultants: Use	No	NS[c]	NS[c]	0.0619
	Yes	NS[c]	NS[c]	(0.999)
Requested Assistance from Wastewater Regulator	No	10	10	0.0238
	Yes	5 (33.33)	4 (28.57)	(1.000)
Compliance Assistance Meets Facility's Environmental Employees' Needs	Lesser: never, some of time, most of time	NS[c]	NS[c]	0.2319
	Greater: always	NS[c]	NS[c]	(0.109)
Importance Management Places on Role of General Plant Employees in Identifying and Correcting Conditions leading to Noncompliance	Lesser: somewhat important, important	NS[c]	NS[c]	0.0714
	Greater: very important	NS[c]	NS[c]	(0.999)

(continued)

TABLE 5.2 (*continued*)

Table 5.2.d. BOD concentration-equivalent limits

Table 5.2.d.i. Qualitative measures of environmental behavior

| Behavioral Measure | Category | Frequency (Subsample Conditional Percent)[a] | | Test Statistic (p-value)[b] |
		Below-Median-Limit Subsample	Above-Median-Limit Subsample	
Management's Steps to Communicate Environmental Performance Goals and Targets to Facility Employees	Below midpoint	NS[c]	NS[c]	0.0851
	At or above midpoint	NS[c]	NS[c]	(0.990)
Management's Steps to Communicate Environmental Performance Goals and Targets to External Parties	Below midpoint	NS[c]	NS[c]	0.0082
	At or above midpoint	NS[c]	NS[c]	(1.000)
ISO 14001 Compliance	No	NS[c]	NS[c]	0.0385
	Yes	NS[c]	NS[c]	(1.000)

[a]The subsample conditional percentage is reported only for the category that represents either the presence of a behavioral measure or the greater extent of a behavioral measure. For example, in the subsample of facilities facing a below-median limit, the reported percentage indicates the number of observations where the facility employs the identified behavior or the greater extent of behavior, while also facing a below-median limit, relative to the number of all facilities facing a below-median limit.

[b]Test statistic applies to the comparison between the below-median-limit subsample and above-median-limit subsample.

[c]NS indicates that at least one cell in the cross-tabulation includes 3 or fewer observations if the cell is populated.

BOD, biological oxygen demand; TSS, total suspended solids.

5.2.d.ii. Quantitative measures of environmental behavior

Behavioral Measure	Limit Subsample	N	Mean	Sign of Difference in Means[a]	Test Statistic (p-value)[b]
Environmental Employees: Absolute	Below median limit	37	100.55	+	-0.189
	Above median limit	37	111.18		(0.850)
Environmental Employees: Relative to Overall Employees	Below median limit	37	0.047	-	0.584
	Above median limit	37	0.034		(0.559)
Wastewater Employees: Absolute	Below median limit	14	10.786	-	-0.730
	Above median limit	13	10.615		(0.465)
Wastewater Employees: Relative to Overall Employees	Below median limit	14	0.122	+	0.388
	Above median limit	13	0.130		(0.698)
Environmental Employees' Education	Below median limit	13	62.615	+	-0.132
	Above median limit	13	66.346		(0.895)
Environmental Employees' Work Experience	Below median limit	14	12.643	-	-0.234
	Above median limit	12	11.667		(0.812)
Training Days	Below median limit	11	7.136	+	0.030
	Above median limit	13	13.269		(0.976)
Training Attendance	Below median limit	12	90.750	+	0.195
	Above median limit	13	92.692		(0.846)
Audit: Annual Count	Below median limit	33	4.242	+	-0.355
	Above median limit	37	13.324		(0.722)
External Consultants: Annual Expenditures—Absolute	Below median limit	12	21,792	+	-1.044
	Above median limit	13	31,538		(0.296)

(continued)

TABLE 5.2 (*continued*)

5.2.d.ii. Quantitative measures of environmental behavior

Behavioral Measure	Limit Subsample	N	Mean	Sign of Difference in Means[a]	Test Statistic (p-value)[b]
External Consultants: Annual Expenditures— Relative to Overall Employees	Below median limit	12	129.54	+	-1.205
	Above median limit	13	360.20		(0.228)
Management's Steps to Communicate Environmental Performance Goals and Targets to Facility Employees (count)	Below median limit	14	3.929	−	-0.076
	Above median limit	13	3.615		(0.940)
Management's Steps to Communicate Environmental Performance Goals and Targets to External Parties (count)	Below median limit	14	2.500	+	0.728
	Above median limit	13	2.769		(0.467)

[a]The difference in mean values = (mean of above median limit subsample) − (mean of below median limit subsample).

[b]Test statistic applies to the comparison between the below-median-limit subsample and above-median-limit subsample; it reports the Wilcoxon two-sample test statistic in the form of a normal approximation.

experience as limits tighten ($\rho = -0.056$), consistent with our hypothesis. Nevertheless, both relationships prove statistically insignificant ($p = 0.425$, $p = 0.661$). Thus, regardless of the dimension of labor quality examined—education or work experience—TSS quantity limits do not appear to influence the labor quality, thus providing no support for our hypothesis that tighter limits induce facilities to employ higher-quality labor effort.

As other measures of labor-effort quality, we explore the effect of TSS production-adjusted quantity limits on the provision of training, its quality, and its quantity. In particular, we explore the effect on the quality of training. As shown in Table 5.2.a, facilities facing tighter limits are much more likely to provide the highest-quality training, consistent with our hypothesis. However, even this large difference proves statistically insignificant.

We also explore the effect of TSS production-adjusted quantity limits on the quantity of training provided. As shown in Table 5.2.a, facilities facing looser limits provide more days of training, based on the mean values, inconsistent with our hypothesis. Yet, the difference in means is very small: 10.2 versus 10.6. (Moreover, the median value is identical in the two subsamples.) In the case of attendance at training sessions, facilities facing looser limits provide training to a greater portion of their environmental employees (as measured by attendance), based on the mean values, inconsistent with our hypothesis. (However, the median value of 100 is identical in the two subsamples.) In both dimensions of training quantity, the difference between facilities facing tighter limits and those facing looser limits proves statistically insignificant. In contrast to the related two-sample test, the corresponding correlation indicates that facilities provide more days of training as limits tighten ($\rho = -0.008$). However, the correlation is statistically insignificant ($p = 0.951$). Similar to the related two-sample test, the corresponding correlation reveals that facilities provide training to a larger portion of their employees as limits loosen ($\rho = 0.018$), inconsistent with our hypothesis, yet this relationship proves statistically insignificant ($p = 0.889$). Thus, regardless of the dimension of training examined—quantity of days or level of attendance—TSS quantity limits do not appear to influence the amount of training, thus providing no support for our hypothesis.

In contrast to labor effort, we explore the effect of TSS production-adjusted quantity limits on the provision of capital equipment. We first explore this effect on the use of any TSS treatment technology and then explore this effect on the use of one of the three more extensive TSS treatment technologies: sedimentation in final clarifier, granular media filtration, and membrane filtration. While we might not expect any relationship, we also explore the effect of TSS production-adjusted quantity limits

on the use of any BOD treatment technology and this effect on the use of
one of the four more extensive BOD treatment technologies: trickling filter biological treatment, activated sludge biological treatment, chemical
oxidation, and carbon absorption. This exploration tests whether the
stringency of discharge limits crosses over from one pollutant to the next.

Consider the effect of the TSS production-adjusted quantity limits on
the presence of any TSS treatment technology. As shown in Table 5.2.a,
facilities facing tighter limits and facilities facing looser limits are nearly
identical in their likelihood of employing TSS treatment. The two-sample
test confirms this point by affirming that the trivial difference is statistically
insignificant. We also assess the effect of TSS limits on the difference between
lesser and greater TSS treatment technologies. As shown, facilities facing
tighter limits are slightly more likely to employ greater TSS treatment. This
slight difference proves statistically insignificant. Thus, regardless of the
categorization of TSS treatment, it is not influenced by TSS limits. We next
assess the effect of TSS production-adjusted quantity limits on BOD treatment. As shown in Table 5.2.a, facilities facing looser TSS limits are somewhat more likely to employ any form of BOD treatment, contrary to either
of our expectations. However, this difference proves statistically insignificant. We also assess the effect of TSS limits on the difference between lesser
and greater BOD treatment technologies. As shown in Table 5.2.a, facilities facing tighter TSS limits are more likely to employ greater BOD treatment. Thus, it appears that regulatory stringency might cross over from
TSS- to BOD-related environmental behavior. However, the noted difference proves statistically insignificant. Thus, regardless of the categorization
of BOD treatment, it is not influenced by TSS limits.

As part of our examination of capital equipment, we also explore the
effect of TSS production-adjusted quantity limits on upgrades to treatment
technologies. We consider first the presence of any upgrade and then only
upgrades designed to reduce wastewater discharges. As shown in Table
5.2.a, facilities facing tighter limits were more likely to perform any upgrade, consistent with our hypothesis. However, this difference does not
prove statistically significant. When considering only upgrades designed to
reduce discharges, facilities facing looser limits are more likely to perform
an upgrade, contrary to our hypothesis. However, this difference proves
statistically insignificant.

Third, we explore the effect of TSS production-adjusted quantity limits on facilities' efforts to understand their environmental management
concerns. In particular, we explore the effect of TSS production-adjusted
quantity limits on the extent of a facility's monitoring program. As shown
in Table 5.2.a, facilities facing tighter limits are more likely to implement
the most extensive monitoring program, consistent with our hypothesis,

though the difference is slight and proves statistically insignificant. Next we explore the influence of TSS production-adjusted quantity limits on the number of audits conducted per year. As shown, facilities facing tighter limits conduct more audits on average, consistent with our hypothesis. Nevertheless, this difference proves statistically insignificant. (Moreover, the median value of one annual audit does not differ between the sub-samples.) The correlation confirms that facilities facing tighter limits conduct more audits per year ($\rho=-0.017$), but this relationship proves statistically insignificant again ($p=0.819$). Thus, regardless of the method for exploring the effect of TSS production-adjusted quantity limits, TSS quantity limits do not appear to influence the per-year count of audits, thus providing no support for our hypothesis.

We also explore the effect of TSS production-adjusted quantity limits on the quality of self-audits, based on the audit team's composition and the extent of the audit classification protocol. As shown in Table 5.2.a, facilities facing tighter limits are more likely to employ the most diverse audit team, consistent with our hypothesis. However, this difference is statistically insignificant. In addition, facilities facing looser limits are more likely to employ the most extensive audit classification protocol, inconsistent with our hypothesis, though the difference is small. Not surprisingly, this difference proves statistically insignificant.

Fourth, we explore the effect of TSS production-adjusted quantity limits on facilities' efforts to seek insight from others regarding their environmental management. We start with facilities' hiring of external consultants, first measured by the presence of this use. As shown in Table 5.2.a, facilities facing looser limits are more likely to hire external consultants, contrary to our hypothesis. However, this difference proves statistically insignificant. We further explore this effect by assessing the amount of money spent on external consultants, in both an absolute sense and relative to a facility's size, as measured by the number of employees. As shown, based on the mean values, facilities facing tighter limits spend more money on consultants in an absolute sense; however, these facilities spend less money relative to their size. The former comparison is consistent with our hypothesis, while the latter comparison is contrary to our hypothesis. Nevertheless, both differences prove statistically insignificant. Correlations support these conclusions. As limits rise, facilities spend less money absolutely but more money relative to their size ($\rho=-0.108$, $\rho=0.033$), with both relationships statistically insignificant ($p=0.395$, $p=0.797$). Thus, regardless of the method for exploring the effect of TSS production-adjusted quantity limits, TSS quantity limits do not appear to influence facilities' expenditures on external consultants, consequently providing no support for our hypothesis.

In addition to external consultants, we explore the effect of TSS production-adjusted quantity limits on facilities' efforts to secure assistance from their regulators. As shown in Table 5.2.a, facilities facing tighter limits are more likely to seek assistance from regulators, consistent with our hypothesis. However, this difference proves statistically insignificant. Similarly, we explore the limits' effect on facilities' success at gathering useful compliance-assistance materials. As shown in Table 5.2.a, facilities facing looser limits are more likely to secure useful compliance assistance, contrary to our hypothesis. However, this difference proves statistically insignificant.

Fifth, we explore the effect of TSS production-adjusted quantity limits on facilities' concern over environmental management and the extent that these concerns translate into goals that are communicated to the facility's employees. In particular, we explore the effect on the level of importance that management places on the role of general plant employees in identifying and correcting conditions that may lead to noncompliance. As shown in Table 5.2.a, facilities facing tighter limits are much more likely to place the highest level of importance on noncompliance, consistent with our hypothesis. However, even this large difference proves statistically insignificant. Similarly, we explore the effect of TSS limits on the degree of effort expended by management to communicate environmental performance goals and targets to facility employees, as measured by whether or not the number of different categories of steps taken by management exceeds the midpoint of three. As shown, facilities facing tighter limits are more likely to take an above-midpoint count of internal communication steps, consistent with our hypothesis. However, this difference proves statistically insignificant. Alternatively, we explore the effect of TSS limits on the actual count of internal communication steps taken. As shown, the count does not differ between the two subsamples to any meaningful degree. The two-sample test confirms that any identifiable difference is statistically insignificant. Similarly, the correlation is very small in magnitude and statistically insignificant ($\rho = 0.063$, $p = 0.618$). Thus, regardless of the assessment method, we conclude that TSS production-adjusted quantity limits do not influence internal communications.

Sixth, in contrast to communication to facility employees, we next explore the effect of TSS production-adjusted quantity limits on communication of environmental performance goals and targets to external parties, as measured by whether or not the number of different categories of steps taken by management exceeds the midpoint of three. As shown in Table 5.2.a, facilities facing tighter limits are more likely to take an above-midpoint count of external communication steps, consistent with our hypothesis, though the difference is small. This small difference proves statistically insignificant. Alternatively, we explore the effect of TSS limits on the actual count of external communication steps taken. As shown, the count does not differ between the two subsamples to any meaningful degree, and the difference is statistically insignificant. Similarly, the correlation is very small in magnitude and statistically insignificant ($\rho = -0.066$, $p = 0.600$). Thus, regardless of the assessment method, we conclude that TSS production-adjusted limits do not influence external communications.

Lastly, we explore the effect of TSS concentration-equivalent limits on the broadest indicator of a well-organized environmental management system: compliance with ISO 14001 standards. As shown in Table 5.2.a, facilities facing tighter limits are more likely to achieve ISO 14001 compliance, consistent with our hypothesis, though the difference is small and proves statistically insignificant.

5.4.1.2. Hypothesis Testing: TSS Concentration-Equivalent Limits. We next consider the effect of TSS concentration-equivalent limits on environmental behavior. First, we examine this effect on the number of environmental employees. As shown in Table 5.2.b, facilities facing tighter limits hire fewer environmental employees—absolutely and relative to overall employees, based on the mean values (and medians), inconsistent with our hypothesis. In both cases, the differences prove statistically significant. Correlations confirm the notion that facilities employ more environmental employees—absolutely and relatively—as their limit levels rise ($\rho = 0.074$, $\rho = 0.305$). However, this relationship proves statistically significant only when the number of environmental employees is scaled relative to the number of overall employees ($p = 0.002$). (For the absolute number, $p = 0.473$.) Thus, tighter TSS concentration limits appear to induce facilities to employ fewer environmental employees, as measured in an absolute sense and relative to overall employees. Both conclusions provide evidence against our hypothesis that tighter limits induce facilities to employ more environmental employees.

Overall, the influence of TSS limits on the number of environmental employees differs between production-adjusted quantity limits and concentration-equivalent limits. While quantity limits do not seem to influence meaningfully facilities' decisions to hire environmental employees, tighter concentration limits appear to induce facilities to hire fewer environmental employees. Thus, the evidence against our hypothesis is incomplete.

Similarly, we explore the effect of TSS concentration-equivalent limits on the number of employees devoted to wastewater management. As shown in Table 5.2.b, facilities facing tighter limits hire more wastewater employees, in an absolute sense, based on the mean values, consistent with our hypothesis. However, facilities facing tighter limits hire fewer wastewater employees, relative to overall employees, based on the mean values, inconsistent with our hypothesis. Regardless of the behavioral measure, the noted differences prove statistically insignificant. Correlations support the conclusions drawn from comparisons of the mean values. As limits rise (loosen), facilities hire fewer wastewater employees in an absolute sense ($\rho = -0.100$) but more wastewater employees relative to overall employees ($\rho = 0.014$). Regardless, these relationships prove statistically insignificant ($p = 0.545$, $p = 0.932$). Thus, tighter TSS concentration limits do not appear to influence facilities' employment of wastewater employees.

Overall, the influence of TSS limits on the number of wastewater employees does not differ between production-adjusted quantity limits and concentration-equivalent limits. Neither quantity nor concentration limits seem to influence meaningfully facilities' decisions to hire wastewater employees.

We next explore the effect of TSS concentration-equivalent limits on the quality of labor effort. As shown in Table 5.2.b, facilities facing looser limits employ a greater proportion of employees with at least a college degree, based on mean values (and medians), contrary to our hypothesis. Also, facilities facing tighter limits employ employees with more work experience, based on mean values, consistent with our hypothesis. However, the difference in mean values is very small. (Moreover, the median is identical in the two subsamples.) In both dimensions of labor quality, the difference between subsamples proves statistically insignificant. Correlations indicate that facilities hire a greater proportion of college-educated environmental employees and workers with greater work experience as limits tighten ($\rho = -0.013$, $\rho = -0.154$), both consistent with our hypothesis. Nevertheless, both relationships prove statistically insignificant ($p = 0.940$, $p = 0.355$) Thus, regardless of the dimension of labor quality examined—education or work experience—TSS concentration limits do not appear to influence the labor quality, thus providing no support for our hypothesis.

Overall, the influence of TSS limits on labor quality does not differ between production-adjusted quantity limits and concentration-equivalent limits. Neither quantity nor concentration limits seem to influence meaningfully facilities' decisions to hire environmental employees of greater quality.

As other measures of labor effort quality, we explore the effect of TSS concentration-equivalent limits on the provision of training, its quality, and its quantity. As shown in Table 5.2.b, facilities facing looser limits are more likely to provide the highest-quality training, inconsistent with our hypothesis, yet this difference is statistically insignificant. Overall, the influence of TSS limits on training quality does not differ between production-adjusted quantity limits and concentration-equivalent limits. Neither quantity nor concentration limits seem to influence meaningfully facilities' decisions to provide training of greater quality.

We also explore the effect of TSS concentration-equivalent limits on the quantity of training. As shown in Table 5.2.b, facilities facing looser limits provide more days of training, based on mean values, inconsistent with our hypothesis. (However, the median is identical in the two sub-samples.) In the case of attendance at training sessions, facilities facing looser limits provide training to a greater portion of their environmental employees, based on mean values, inconsistent with our hypothesis. (However, the median is identical in the two subsamples.) In both dimensions of training, the differences between subsamples prove statistically insignificant. In contrast to the related two-sample test, the corresponding correlation indicates that facilities provide more days of training as limits tighten ($\rho = -0.012$), consistent with our hypothesis. However, the relationship is statistically insignificant ($p = 0.946$). Similar to the related two-sample test, the corresponding correlation reveals that facilities provide training to a larger portion of their employees as limits loosen ($\rho = 0.160$), inconsistent with our hypothesis. Nevertheless, this relationship proves statistically insignificant ($p = 0.345$). Thus, regardless of the dimension examined, TSS concentration limits do not appear to influence the amount of training, thus providing no support for our hypothesis.

Overall, the influence of TSS limits on training quantity does not differ between production-adjusted quantity limits and concentration-equivalent limits. Neither quantity nor concentration limits seem to influence meaningfully facilities' decisions to provide more training.

Next, we explore the effects of TSS concentration-equivalent limits on the use of TSS treatment technologies and BOD treatment technologies in order. As shown in Table 5.2.b, facilities facing looser TSS limits are more likely than facilities facing tighter limits to employ some form of TSS treatment (in this case, the former set of facilities are fully likely), contrary to our hypothesis. However, this difference proves statistically insignificant. We also assess the effect of TSS concentration-equivalent limits on the distinction between lesser and greater TSS treatment. As shown, similar to the distinction between any TSS treatment and none, facilities facing looser limits are more likely to employ greater TSS treatment, contrary to our hypothesis. Yet, again this difference proves statistically insignificant. Thus, regardless of the categorization of TSS treatment, it is not influenced by TSS limits. As a test of whether the effect of discharge limits crosses over from one pollutant to the next, we assess the influence of TSS limits on BOD treatment. As with TSS treatment, facilities facing looser TSS limits are more likely to employ BOD treatment, contrary to the notion that limits might cross over from one pollutant to the next. Yet, the difference proves insignificant, consistent with our initial expectation that the limit on one pollutant does not influence the treatment of another pollutant. An identical conclusion stems from our assessment of the distinction between lesser and greater BOD treatment. Thus, regardless of the categorization of BOD treatment, it is not influenced by TSS limits.

Overall, neither TSS production-adjusted quantity limits nor TSS concentration-equivalent limits influence either TSS treatment or BOD treatment, providing no support for the expectation that tighter limits prompt better treatment.

As part of our examination of capital equipment, we also explore the effect of TSS concentration-equivalent limits on upgrades to treatment technologies. We first consider the presence of any upgrade and then only upgrades designed to reduce wastewater discharges. As shown in Table 5.2.b, facilities facing tighter limits are more likely to perform an upgrade, consistent with our hypothesis. However, this difference does not prove statistically significant. Similarly, facilities facing tighter limits are more likely to perform an upgrade designed to reduce discharges, consistent with our hypothesis. However, again this difference proves statistically insignificant.

Overall, the influence of TSS limits on treatment upgrades does not differ between production-adjusted quantity limits and concentration-equivalent limits. Neither quantity nor concentration limits seem to influence meaningfully facilities' decisions to upgrade their treatment equipment, regardless of the intent of reducing discharges.

Third, we explore the effect of TSS concentration-equivalent limits on facilities' efforts to understand their environmental management concerns. As shown in Table 5.2.b, facilities facing looser limits are more likely to implement the most extensive monitoring program, inconsistent with our hypothesis, though the difference is slight and statistically insignificant. Overall, the influence of TSS limits on monitoring programs does not differ between production-adjusted quantity limits and concentration-equivalent limits. Neither quantity nor concentration limits seem to influence meaningfully the extent of a facility's monitoring program.

Next we explore the influence of TSS concentration-equivalent limits on the number of audits conducted per year. As shown, facilities facing tighter limits conduct significantly more audits, based on mean values (and medians), consistent with our hypothesis. The related correlation confirms the notion that facilities conduct more audits as TSS concentration-equivalent limits tighten ($\rho = -0.099$), but this relationship is statistically insignificant ($p = 0.359$). Thus, tighter TSS concentration limits appear to induce facilities to conduct more audits per year (at least based on the two-sample test), which provides evidence for our hypothesis.

Overall, the influence of TSS limits on the per-year count of audits differs between production-adjusted quantity limits and concentration-equivalent limits. While quantity limits do not seem to meaningfully influence facilities' decisions to conduct audits, tighter concentration limits appear to induce facilities to conduct more audits per year, thus providing at least some support for our hypothesis that tighter limits induce facilities to conduct more audits per year.

We also explore the effect of TSS concentration-equivalent limits on the quality of self-audits. As shown in Table 5.2.b, facilities facing looser limits are more likely to employ the most diverse audit team, inconsistent with our hypothesis, yet this difference is statistically insignificant. Similarly, facilities facing looser limits are more likely to employ the most extensive audit classification protocol, inconsistent with our hypothesis, yet this difference proves statistically insignificant.

Overall, the influence of TSS limits on audit quality does not differ between production-adjusted quantity limits and concentration-equivalent limits. Neither quantity nor concentration limits seem to meaningfully influence either dimension of audit quality: the composition of a facility's audit team or the extent of the audit classification protocol.

Fourth, we explore the effect of TSS concentration-equivalent limits on facilities' efforts to seek insight from others regarding their environmental management. We start with facilities' hiring of external consultants. As shown in Table 5.2.b, facilities facing tighter limits are more likely to hire external consultants, consistent with our hypothesis. However, this difference is small (81 percent versus 77 percent) and proves statistically insignificant. We further explore this effect by assessing the amount of money spent on external consultants, in both an absolute sense and relative to a facility's size. Based on mean values, facilities facing looser limits spend more money on consultants in an absolute sense and relative to their size. These comparisons are inconsistent with our hypothesis, yet both prove statistically insignificant. Correlations do not support these conclusions. As limits rise, facilities spend less money absolutely but more money relative to their size ($\rho = -0.050$, $\rho = 0.218$). Still, consistent with the two-sample tests, both relationships prove statistically insignificant ($p = 0.769$, $p = 0.200$). Thus, regardless of the method for exploring the effect of TSS concentration-equivalent limits, TSS concentration limits do not appear to influence facilities' use of external consultants, consequently providing no support for our hypothesis.

Overall, the influence of TSS limits does not differ between production-adjusted quantity limits and concentration-equivalent limits. Neither quantity nor concentration limits seem to meaningfully influence facilities' decisions to use external consultants.

In addition to external consultants, we explore the effect of TSS concentration-equivalent limits on facilities' efforts to secure assistance from their regulators. As shown in Table 5.2.b, facilities facing tighter limits are much more likely to seek assistance from regulators, strongly consistent with our hypothesis and statistically significant. As part of this comparison, none of the facilities facing looser limits seek assistance from their regulators.

Overall, the influence of TSS limits differs between production-adjusted quantity limits and concentration-equivalent limits. While the former limit type does not influence efforts to seek regulatory assistance, the latter limit type appears to induce facilities to seek regulatory assistance more when facing tighter limits.

Next, we explore the limits' effect on facilities' success at gathering useful compliance-assistance materials. As shown in Table 5.2.b, facilities facing looser limits are more likely to secure useful compliance assistance, contrary to our hypothesis, yet this difference proves statistically insignificant. Overall, neither TSS production-adjusted quantity nor TSS concentration-equivalent limits appear to influence facilities' success at gathering useful compliance-assistance materials.

Fifth, we explore the effect of TSS concentration-equivalent limits on facilities' concern over environmental management and the extent that these concerns translate into goals that are communicated to facilities' employees. As shown in Table 5.2.b, facilities facing looser limits are more likely to place the highest level of importance on the role of general plant employees in identifying and correcting conditions that may lead to noncompliance, inconsistent with our hypothesis, though the difference is small and statistically insignificant.

Overall, the influence of TSS limits does not differ between production-adjusted quantity limits and concentration-equivalent limits. Neither quantity nor concentration limits seem to meaningfully influence facilities' concern over environmental management.

Similarly, we explore the effect of TSS concentration-equivalent limits on the degree of effort expended by management to communicate environmental performance goals and targets to facility employees, as measured by whether or not the number of different categories of steps taken by management exceeds the midpoint of three. As shown in Table 5.2.b, facilities facing tighter limits are more likely to take an above-midpoint count of internal communication steps, consistent with our hypothesis, though the difference is small and statistically insignificant. Alternatively, we explore the effect of TSS concentration-equivalent limits on the actual count of internal communication steps taken. As shown, the count does not differ greatly between the two subsamples, and the difference is statistically insignificant. Similarly, the correlation is very small in magnitude and statistically insignificant ($\rho = -0.047$, $p = 0.776$). Thus, we draw the same conclusion: TSS concentration-equivalent limits do not influence the extent of internal communications, providing no support for our hypothesis that tighter limits prompt facilities' management to better internally communicate its environmental performance goals and targets.

Overall, the influence of TSS limits does not differ between production-adjusted quantity limits and concentration-equivalent limits. Neither quantity nor concentration limits seem to meaningfully influence the degree of effort expended by management to communicate to facility employees.

Sixth, in contrast to the communication to facility employees, we next explore the effect of TSS concentration-equivalent limits on the communication of environmental performance goals and targets to external parties, as measured by whether or not the number of different categories of steps taken by management exceeds the midpoint of three. As shown in Table 5.2.b, facilities facing looser limits are more likely to take an above-midpoint count of external communication steps, inconsistent with our hypothesis, yet this difference proves statistically insignificant. Alternatively, we explore the effect of TSS concentration-equivalent limits on the actual count of external communication steps taken. As shown, facilities facing looser limits take more steps to communicate their environmental performance goals to external parties, but this difference is statistically insignificant. Similarly, the correlation is small in magnitude and statistically insignificant ($\rho = -0.098$, $p = 0.551$). Overall, the influence of TSS limits does not differ between production-adjusted quantity limits and concentration-equivalent limits. Neither quantity nor concentration limits seem to meaningfully influence facilities' communication to external parties.

Lastly, we explore the effect of TSS concentration-equivalent limits on compliance with ISO 14001 standards. As shown in Table 5.2.b, facilities facing looser limits are more likely to achieve ISO 14001 compliance, inconsistent with our hypothesis, though the difference is small and statistically insignificant. Overall, the influence of TSS limits does not differ between production-adjusted quantity limits and concentration-equivalent limits. Neither quantity nor concentration limits seem to meaningfully influence compliance with ISO 14001 standards.

5.4.2. Hypothesis Testing: BOD Discharge Limits

We next explore the effects of BOD limits on the various forms of environmental behavior. Rather than exhaustively exploring each behavioral form, we instead focus on those results that indicate meaningful relationships between BOD limits and environmental behavior, that is, those relationships for which the relevant two-sample tests and correlation statistics prove statistically significant. We explore first the effects of BOD production-adjusted quantity limits and second the effects of BOD concentration-equivalent limits. As part of our exploration of concentration-equivalent limits, we attempt to summarize the effect of BOD limits in general on each behavioral form.

5.4.2.1. Hypothesis Testing: BOD Quantity Limits. First, we consider the effect of BOD production-adjusted quantity limits on the number of environmental employees. Based on the two-sample tests and correlations, BOD production-adjusted quantity limits do not appear to influence either the absolute or the relative number of environmental employees facilities choose (respectively, correlations equal 0.062 and 0.019, with $p = 0.417$ and $p = 0.811$). We next explore the effect of BOD production-adjusted quantity limits on the number of wastewater-related employees. As shown in Table 5.2.c, the difference in the absolute number of wastewater-related employees between facilities facing tighter limits and facilities facing looser limits proves statistically insignificant. Yet, facilities facing tighter limits hire significantly more wastewater employees, relative to overall employees, consistent with our hypothesis.[8] The correlations indicate that as limit levels fall (limits tighten), facilities employ fewer wastewater employees absolutely ($\rho = 0.215$) but more wastewater employees relative to the number of overall employees ($\rho = -0.059$). The former result is inconsistent with our hypothesis and proves statistically significant ($p = 0.105$), while the latter result is consistent with our hypothesis but proves statistically insignificant ($p = 0.665$). Thus, our two methods for exploring the effect of TSS production-adjusted quantity limits on the number of wastewater employees and the two manners of measuring wastewater employees (absolute versus relative) generate different conclusions.

BOD production-adjusted quantity limits fail to influence the quality of labor effort as measured by the education and work experience of environmental employees, based on both two-sample tests and correlations (respectively, correlations are $\rho = 0.191$ and $\rho = -0.062$, with significance levels of $p = 0.156$ and $p = 0.647$).

Similarly, BOD production-adjusted quantity limits do not significantly influence the quality of training provided by facilities. Moreover, BOD production-adjusted quantity limits do not significantly influence the quantity of training provided by facilities in terms of days or attendance, regardless of the method used to explore the relationship (respectively, correlations are $\rho = 0.021$ and $\rho = 0.038$, with significance levels of $p = 0.882$ and $p = 0.779$).

Next, we explore the effects of BOD production-adjusted quantity limits on the use of BOD treatment technologies and TSS treatment technologies in order. As shown in Table 5.2.c, facilities facing looser BOD limits are more likely than facilities facing tighter limits to employ some form of BOD treatment, contrary to our hypothesis, though the difference is small. Not surprising, this difference proves statistically insignificant. Unlike the distinction between any BOD treatment and none, facilities facing tighter limits are more likely to employ greater BOD treatment, consistent with our hypothesis. However, this difference proves statistically insignificant. Thus, regardless of the categorization of BOD treatment, it is not influenced by BOD limits.

As a test of whether the effect of discharge limits crosses over from one pollutant to the next, we assess the influence of BOD limits on TSS treatment. As shown in Table 5.2.c, while facilities facing looser BOD limits are more likely to employ TSS treatment, the difference proves statistically insignificant. Thus, the limit on one pollutant does not seem to influence the treatment of another pollutant. An identical conclusion stems from our assessment of the distinction between lesser and greater TSS treatment. Thus, regardless of the categorization of TSS treatment, it is not influenced by BOD limits.

As for treatment upgrades, BOD production-adjusted quantity limits do not significantly influence the facilities' decisions to upgrade their treatment operations, regardless of whether or not these upgrades are designed to reduce wastewater discharges.

BOD production-adjusted quantity limits do not significantly influence the extent of a facility's monitoring program. Similar to the extent of a facility's monitoring program, BOD production-adjusted quantity limits do not appear to influence the count of audits conducted per year by facilities, based on the two-sample tests and the correlation ($\rho = -0.0592$, $p = 0.448$). Similar to the count of audits, BOD production-adjusted quantity limits do not appear to influence the quality of self-audits conducted by facilities, measured by either the composition of the audit team or the rigor of the classification protocol.

Similar to monitoring and auditing efforts, efforts to secure information from external sources, in particular from consultants measured by either the presence of consultant use or expenditures on consultants, do not seem influenced by BOD production-adjusted quantity limits, based on both two-sample tests and correlations. (Correlations for absolute expenditures and relative expenditures are $\rho=-0.102$ and $\rho=0.0723$, respectively, with significance levels of $p=0.447$ and $p=0.590$).

Similar to the use of external consultants, facilities' efforts to seek assistance from their regulators do not seem influenced by BOD production-adjusted quantity limits. As shown in Table 5.2.c, an identical percentage of facilities seek regulatory assistance in the two subsamples; that is, facilities facing tighter BOD production-adjusted quantity limits are as likely to seek regulatory assistance as are facilities facing looser limits. Moreover, BOD production-adjusted quantity limits do not appear to influence facilities' success at securing useful compliance-assistance materials. In addition, BOD production-adjusted quantity limits do not appear to influence managements' concerns over noncompliance. The lack of influence of BOD production-adjusted quantity limits on managements' concerns over environmental matters extends to the degree of effort expended by management to communicate environmental performance goals and targets to facility employees. Regardless of the method for exploring the managers' internal communication efforts, BOD production-adjusted quantity limits do not significantly influence these efforts. (For the count of communication methods, the correlation is $\rho=0.099$, with $p=0.459$). Similar to internal communication efforts, the communication of environmental performance goals and targets to external parties does not appear influenced by BOD production-adjusted quantity limits, regardless of the method for exploring external communication efforts. (For the count of communication methods, the correlation is $\rho=-0.023$, with $p=0.867$).

Lastly, BOD production-adjusted quantity limits do not significantly influence facilities' compliance with ISO 14001 standards.

5.4.2.2. Hypothesis Testing: BOD Concentration-Equivalent Limits. Finally, we explore the effects of BOD concentration-equivalent limits on the various forms of environmental behavior. Table 5.2.d provides the relevant statistics. For each behavioral form, we summarize the influence of BOD limits in general based on the effects of both production-adjusted quantity and concentration-equivalent limits. In the end, we attempt to summarize the influence of discharge limits in general based on the effects of both TSS and BOD limits.

First, we consider the effect of BOD concentration-equivalent limits on the number of environmental employees. Based on the two-sample tests shown in Table 5.2.d, BOD concentration-equivalent limits do not seem to affect the number of environmental employees, regardless of measurement form (absolute or relative to the number of overall employees). In contrast, the correlations reveal that facilities employ fewer environmental employees, in an absolute sense, as the BOD concentration-equivalent limit tightens (ρ = 0.416, p = 0.000), inconsistent with our hypothesis. (Based on the ratio of environmental to overall employees, the correlation is ρ = 0.080, with a significance level of p = 0.497). Overall, BOD limit results provide some evidence against our hypothesis by indicating that tighter BOD concentration-equivalent limits appear to induce facilities to hire fewer environmental employees in an absolute sense.

Considering both TSS and BOD limits, tighter TSS limits (in concentration-equivalent form) and tighter BOD limits (in concentration-equivalent form) appear to induce facilities to hire fewer environmental employees, providing two pieces of evidence against our hypothesis that tighter limits induce facilities to hire more environmental employees.

We next explore the effect of BOD concentration-equivalent limits on the number of wastewater-related employees. The two-sample tests reveal no significant difference in wastewater employees—absolutely or relative to overall employees—between facilities facing tighter limits and facilities facing looser limits. However, correlations reveal that as limit levels loosen, facilities employ significantly more wastewater employees absolutely (ρ = 0.390, p = 0.044). This result is inconsistent with our hypothesis. (Regarding the relative number of wastewater employees, the correlation of ρ = 0.043 proves statistically insignificant, given a significance level of p = 0.832.) Thus, our two methods for exploring the effect of BOD concentration-equivalent limits on the absolute number of wastewater employees generate different conclusions. Overall, BOD limit results provide some evidence against our hypothesis by indicating that tighter BOD limits—both quantity and concentration limits—induce facilities to hire fewer wastewater employees, especially in an absolute sense.

Considering both TSS and BOD limits, the evidence against our hypothesis is substantial but incomplete. Tighter BOD limits—both quantity limits and concentration limits—induce facilities to hire fewer wastewater employees. However, neither TSS production-adjusted quantity limits nor TSS concentration-equivalent limits seem to meaningfully influence facilities' decisions to hire wastewater employees.

In contrast to labor quantity, BOD concentration-equivalent limits fail to influence the quality of labor effort as measured by the education and work experience of environmental employees, based on both two-sample tests and correlations (respectively, the correlations are ρ = 0.299 and ρ = −0.106, with significance levels of p = 0.138 and p = 0.605). Overall, both types of BOD limits fail to influence the quality of labor effort. Considering both TSS and BOD limits, discharge limits do not seem to meaningfully influence facilities' decisions to hire environmental employees of greater quality.

Similarly, BOD production-adjusted quantity limits do not significantly influence the quality of training provided by facilities. Overall, neither types of BOD limits—production-adjusted quantity nor concentration-equivalent—influence the quality of training provided by facilities. Considering both TSS and BOD limits, discharge limits do not seem to meaningfully influence facilities' decisions to provide training of greater quality.

Moreover, BOD production-adjusted quantity limits do not significantly influence the quantity of training provided by facilities. (Correlations associated with days of training and attendance at sessions are $\rho = 0.225$ and $\rho = 0.135$, respectively, with significance levels of $p = 0.291$ and $p = 0.521$.) Overall, regardless of the dimension of training examined—quantity of days or level of attendance—and the analytical method employed, BOD limits do not appear to influence the amount of training, thus providing no support for our hypothesis that tighter limits induce facilities to employ more training. Considering both TSS and BOD limits, discharge limits do not seem to meaningfully influence facilities' decisions to provide training.

Next, we explore the effects of BOD concentration-equivalent limits on the use of BOD treatment technologies and TSS treatment technologies. As shown in Table 5.2.d, facilities facing looser BOD limits are more likely than facilities facing tighter limits to employ some form of BOD treatment, contrary to our hypothesis. However, this difference proves statistically insignificant. Similarly, facilities facing looser limits are more likely to employ greater BOD treatment, inconsistent with our hypothesis, but insignificantly so. Thus, regardless of the categorization of BOD treatment, it is not influenced by BOD limits. Contrary to BOD treatment, facilities facing tighter BOD limits are more likely to employ TSS treatment, consistent with the notion that limits cross over from one pollutant to the next. As shown in Table 5.2.d, *all* facilities facing tighter BOD limits employ some form of TSS treatment. Despite this point, the difference between the two subsamples proves insignificant, consistent with our initial expectation that the limit on one pollutant does not influence the treatment of another pollutant. Results from an assessment of the distinction between lesser and greater TSS treatment also reveal an insignificant difference. Thus, regardless of the categorization of TSS treatment, it is not influenced by BOD limits. Overall, neither BOD production-adjusted quantity limits nor BOD concentration-equivalent limits influence either BOD treatment or TSS treatment. Considering both TSS limits and BOD limits, discharge limits do not influence either TSS treatment or BOD treatment, providing no support for the expectation that tighter limits prompt better treatment.

BOD concentration-equivalent limits do not significantly influence facilities' decisions to upgrade their treatment operations, regardless of whether or not these upgrades are designed to reduce wastewater discharges. Overall, neither BOD production-adjusted quantity limits nor BOD concentration-equivalent limits seem to influence facilities' decisions to upgrade treatment, regardless of whether these upgrades are designed to reduce discharges. Considering both TSS and BOD limits, discharge limits do not seem to meaningfully influence facilities' decisions to upgrade their treatment operations.

Beyond treatment, BOD concentration-equivalent limits do not significantly influence the extent of a facility's monitoring program. Overall, the influence of BOD limits on monitoring programs does not differ between production-adjusted quantity limits and concentration-equivalent limits. Neither quantity nor concentration BOD limits seem to meaningfully influence the extent of a facility's monitoring program. Considering both TSS and BOD limits, discharge limits do not affect the extent of a facility's monitoring program.

Next, we consider the effect of BOD concentration-equivalent limits on the count of audits conducted per year by facilities. Based on the two-sample test results shown in Table 5.2.d, BOD concentration-equivalent limits do not seem to affect the count of audits. In contrast, the correlation reveals that facilities conduct significantly fewer audits as the BOD concentration-equivalent limit tightens ($\rho = 0.205$, $p = 0.090$). Overall, BOD limit results provide some evidence against our hypothesis by indicating that at least tighter BOD concentration-equivalent limits induce facilities to conduct fewer audits; in contrast, BOD production-adjusted quantity limits do not appear to influence the count of audits conducted. Considering both TSS and BOD limits, the results are contradictory. Our results reveal that tighter TSS limits (in concentration-equivalent form) appear to induce facilities to conduct more audits per year, yet tighter BOD limits (in concentration-equivalent form) appear to induce facilities to conduct fewer audits. We are at a loss to explain these contradictory conclusions.

In contrast to the quantity of audits, BOD concentration-equivalent limits do not appear to influence the quality of self-audits conducted by facilities as measured by either the composition of the audit team or the rigor of the audit classification protocol. Overall, the influence of BOD limits on audit quality does not differ between production-adjusted quantity limits and concentration-equivalent limits: neither quantity nor concentration BOD limits seem to meaningfully influence audit quality. Considering both TSS and BOD limits, discharge limits do not affect the quality of self-audits conducted by facilities.

Still, BOD concentration-equivalent limits seem to influence efforts to secure information from external sources by hiring consultants as measured by expenditures on consultants (when scaled by facility size), at least based on the correlation ($\rho = 0.514$, $p = 0.009$): as limits rise, expenditures increase, inconsistent with our hypothesis. The two-sample test does not support this conclusion. Similarly, without scaling, consultant expenditures do not seem influenced by discharge limits, based on the two-sample test and correlation ($\rho = 0.113$, $p = 0.590$). The same conclusion applies to the presence of consultants. Overall, the influence of BOD limits differs between production-adjusted quantity limits and concentration-equivalent limits. Only the latter limits seem to meaningfully influence facilities' decisions to use external consultants. Considering both TSS and BOD limits, only BOD discharge limits seem to influence the use of consultants, while providing evidence against our hypothesis that tighter limits induce facilities to seek assistance from external sources such as consultants.

Unlike the use of consultants, BOD concentration-equivalent limits do not seem influence facilities' efforts to seek assistance from their regulators. Overall, the influence of BOD limits does not differ between production-adjusted quantity limits and concentration-equivalent limits. Regardless of the limit type, BOD limits do not influence facilities' efforts to seek regulatory assistance. Considering both TSS and BOD limits, the influence of discharge limits differs between the two pollutants. While the former pollutant limit (in concentration-equivalent form) appears to induce facilities to seek regulatory assistance more when facing tighter limits, the latter pollutant limit does not influence facilities' efforts to seek regulatory assistance. Thus, these results provide only partial support for our hypothesis that tighter limits induce facilities to seek assistance from external sources such as regulators.

We next explore the influence of BOD concentration-equivalent limits on facilities' success at securing useful compliance-assistance materials. As shown in Table 5.2.d, facilities facing tighter BOD limits are more likely to provide useful compliance-assistance materials, and this difference is statistically significant. Overall, the influence of BOD limits differs between production-adjusted quantity limits and concentration-equivalent limits. While the former limits do not appear to influence facilities' success at gathering useful compliance-assistance materials, the latter limits appear to drive facilities to secure useful compliance-assistance materials when facing tighter limits. Considering both TSS and BOD limits, the influence of discharge limits differs between the two pollutants. While the former pollutant limit does not influence facilities' success at securing useful compliance-assistance materials, the latter pollutant limit (in concentration-equivalent form) appears to induce facilities to secure useful compliance-assistance materials when facing tighter limits. Thus, these results provide only partial support for our hypothesis that tighter limits induce facilities to gather information about their environmental management issues.

In contrast, BOD concentration-equivalent limits do not appear to influence managements' concerns over noncompliance. Overall, neither BOD production-adjusted quantity limits nor BOD concentration-equivalent limits seem to meaningfully influence facilities' concern over environmental management. Considering both TSS and BOD limits, discharge limits do not meaningfully affect facilities' concern.

Similarly, BOD concentration-equivalent limits do not significantly influence the degree of effort expended by management to communicate environmental performance goals and targets to facility employees, regardless of the analytical method employed (based on the underlying count of communication steps, the correlation is $\rho = -0.280$, with a significance level of $p = 0.157$). Overall, neither BOD production-adjusted quantity limits nor BOD concentration-equivalent limits seem to meaningfully influence the degree of effort expended by management to communicate environmental performance goals and targets to facility employees. Considering both TSS and BOD limits, discharge limits do not meaningfully affect managers' internal communication efforts, providing no support for our hypothesis that tighter discharge limits prompt facilities to communicate better to facility employees.

Similar to internal communication efforts, the communication of environmental performance goals and targets to external parties does not appear influenced by BOD concentration-equivalent limits, regardless of the method for exploring efforts (based on the underlying count of communication steps, the correlation is $\rho = -0.122$, with a significance level of $p = 0.545$). Overall, neither BOD production-adjusted quantity limits nor BOD concentration-equivalent limits seem to meaningfully influence facilities' external communication efforts. Considering both TSS and BOD limits, discharge limits do not meaningfully affect facilities' external communication, providing no support for our hypothesis that tighter discharge limits prompt facilities to communicate better externally.

Lastly, BOD concentration-equivalent limits do not significantly influence facilities' compliance with ISO 14001 standards. Overall, neither BOD production-adjusted quantity limits nor BOD concentration-equivalent limits seem to meaningfully influence compliance with ISO 14001 standards. Considering both TSS and BOD limits, discharge limits do not meaningfully affect facilities' compliance with ISO 14001 standards, providing no support for our hypothesis that tighter discharge limits prompt facilities to improve their environmental management practices comprehensively, in this case measured by ISO 14001 compliance.

5.5. SUMMARY

Our principal inquiry in this chapter is how imposed discharge limits affect environmental behavior by regulated facilities in the chemical manufacturing industry. We test the hypothesis that tighter limits prompt better environmental behavior. Among the behaviors measured and analyzed are quantity and quality of labor effort, provision of capital equipment, facility monitoring, the use of consultants, communication of environmental goals to facility employees and external employees, and compliance with ISO 14001 standards. The findings in this chapter indicate that for most of the forms of behavior, which we would expect to generate better environmental performance, our empirical results reveal no link from discharge limits to these presumably desirable forms of environmental behavior. These results may be influenced by several different factors. First, the size of our sample may not be large enough to generate statistically significant results. Second, we may not have accurately identified the forms of behavior that are most likely to prompt desirable changes in environmental performance. Third, facilities facing tightened discharge limits may seek ways to cut costs to compensate for the increased costs they anticipate will result from discharge limits that are more costly and difficult to achieve. Thus, for example, a facility that will incur higher treatment costs may decide to cut its payroll (by laying off employees) to prevent a decrease in profits, even though some of those employees may be environmental employees who might be useful in achieving compliance with the tightened limits. In this sense, a tightening of a discharge limit may generate an apparently counterproductive effect. Fourth, a tightening of a discharge limit may prompt recalcitrance from a regulated facility as it faces what it regards as a regulatory obligation that is evermore difficult to achieve. Fifth, the facilities in our sample may have believed during the relevant period that enforcement of tighter discharge limits was unlikely or that, even if enforcement were to occur, any resulting sanctions would not be severe enough to justify before-the-fact changes in environmental behavior.

Chapter Five Appendix

This first section of the Chapter 5 appendix assesses whether and to what degree environmental behavior varies across a variety of dimensions. Facilities operating in different states or EPA regions may differ systematically in the extent of their environmental management efforts. Similarly, facilities operating in different sectors may differ systematically in their environmental management efforts. As important, larger facilities may employ environmental management strategies systematically different from those of smaller facilities. We also explore differences over time; certain forms of environmental management may show a positive trend, while others show a negative trend. As a complement to an assessment of overall trends, we also explore trends within individual facilities by analyzing variation over time for each specific facility and summarizing these individual trends across all of the sampled facilities. For these assessments across states, regions, sectors, and facility size categories, as with the hypothesis testing described in Chapter 5, we simplify our analysis of environmental management practices that we measure in ordered categories or as a count.

5A.1.1. Subsamples Based on Location: EPA Regions

As the first set of subsamples, we explore variation across facility location. In particular, we assess differences in environmental management practices across EPA regions. Rather than exploring each region separately, we identify the two most prominent regions in our sample—Region 4 and Region 6—while grouping the remaining regions into a single "other region" category. This variation reveals interesting patterns. For example, examination of the presence versus the absence of any BOD treatment technology reveals that the presence of BOD treatment is most likely in Region 6: 90 percent of the facilities in this region operate a BOD treatment technology; in contrast, the percentages in Region 4 and other regions are 79 percent and 76 percent, respectively. Similarly, an assessment of the distinction between greater and lesser BOD treatment technologies reveals that the use of greater technologies is most prevalent in Region 6 (61 percent of facilities in this region as opposed to 51 percent in other regions), but the difference between Regions 4 and 6 is very small (<1 percent). Regarding TSS treatment technologies, an examination of the presence versus the absence reveals that presence does not vary across regions much at all: the percentage of facilities operating a TSS treatment technology equals 89 percent, 87 percent, and 89 percent, respectively, for Regions 4, 6, and other. However, an assessment of the distinction between greater and lesser TSS treatment technologies reveals that the use of greater technologies is substantially more prevalent in Regions 4 and 6 (61 percent and 58 percent) as opposed to other regions (39 percent). We neither report not interpret all of the results of exploration. Instead, we offer a full tabular display of the regional variation on the Web site of Stanford University Press. This display excludes cases where individual facilities may be revealed; in particular, we do not display a cross-tabulation between behavior and EPA region if any cell of the cross-tabulation contains three or fewer facilities.

5A.1.2. Subsamples Based on Chemical Manufacturing Subsectors

In this section, we assess differences in environmental management practices across the three broad sectors: organic chemicals, inorganic chemicals, and other chemicals. The full tabular display of sectoral variation is provided on the Web site of Stanford University Press.

As part of this assessment, we test whether any noted difference between the organic chemical manufacturing subsector and the inorganic chemical manufacturing subsector is statistically significant. In essence, we test whether the prevalence of a given management practice differs significantly between these two subsectors. When measuring a particular management practice that falls into two categories (for example, ISO compliance presence or absence), we employ a two-sample test. When measuring a particular management practice across a spectrum of values (for example, number of facility-level environmental employees), we employ two two-sample tests: Wilcoxon two-sample test and median two-sample test.[1] In the tabular display, we report only the former test results since the latter test results lead to identical conclusions (except environmental employees' work experience, where the Wilcoxon test and median test p-values equal 0.095 and 0.159, respectively.)

As the first part of our assessment, we explore whether the quantity of labor devoted to environmental management differs across the three broad sectors. Based on the median values, organic chemical facilities employ a larger absolute number of environmental employees than do inorganic chemical facilities (60 versus 47), while the median values do not differ between organic chemicals facilities and other chemical facilities (60). Based on the mean values, organic chemical facilities again employ a larger absolute number of environmental employees than do inorganic chemical facilities (108 versus 72) but a smaller absolute number of environmental employees than do other chemical facilities (108 versus 133). Relative to the number of overall employees, the number of environmental employees differs little across the three broad sectors, based on both median values and mean values. Two-sample tests indicate that organic chemical facilities employ significantly more environmental employees than do inorganic chemical facilities (p = 0.076). However, as perhaps the better measure, the relative number of environmental employees does not differ significantly between these two primary sectors (p = 0.843).

Within the category of environmental employees, we focus on the use of wastewater employees and assess whether the number of wastewater employees varies across the three broad sectors. Considering the absolute number, organic chemical facilities employ more wastewater employees than do inorganic chemical facilities, based on both median and mean values. Relative to the number of overall employees, the same comparison holds true based on the median values, though the difference is small (0.034 versus 0.032), but the opposite comparison holds based on the mean values, and the difference is substantial (0.136 versus 0.062). Nevertheless, neither the absolute number of wastewater employees nor the relative number of wastewater employees significantly differs between inorganic chemical facilities and organic chemical facilities (respectively, p = 0.166 and p = 0.940).

We next explore how the educational attainment of environmental employees differs across the three broad sectors. Organic chemical facilities employ the largest share of college-educated environmental employees, based on both the median and mean values (respectively, 67 and 80). Based on the two-sample test, this difference in the level of education between inorganic chemical facilities and organic chemical facilities is statistically significant (p = 0.050). The work experience of environmental employees may also differ across the broad sectors. Inorganic chemical facilities use environmental employees with the most work experience, based on both the median and mean values (respectively, 14.8 and 15). The two-sample test indicates that the degree of work experience differs significantly between inorganic chemical facilities and organic chemical facilities (p=0.095).

We next explore the quality of training provided to environmental employees. The prevalence of the highest-quality training is strongest in the inorganic chemical manufacturing sector (13 of 28 facilities). However, based on the two-sample test, the prevalence of this type of training does not significantly differ between the two primary broad sectors (p=0.999). We also explore the quantity of training, as measured by the number of training days, which does not differ across the broad sectors based on the median value (5). Based on the mean value, facilities operating in the inorganic chemical manufacturing sector provide the most training (12.3). However, no significant difference exists between the two primary broad sectors (p=0.179). We also measure the quantity of training based on the proportion of environmental employees who generally attend any provided training. The median facility in each subsample provides training to all of its environmental employees. The mean facility in the other-chemical broad sector provides training to the largest proportion of its environmental employees (95.9), and the mean inorganic chemical facility provides training to a larger proportion than does the mean organic chemical facility (90.3 versus 87.8). However, the level of training attendance does not significantly differ between the two primary broad sectors (p=0.591).

In contrast to labor effort, we next assess facilities' use of capital equipment. First, we consider TSS treatment technologies. Initially, we assess the presence versus the absence of any TSS treatment technology. Organic chemical facilities are slightly more likely to employ a TSS treatment technology (44 of 50 facilities). However, this slight difference proves statistically insignificant (p=0.999). We also assess the distinction between greater and lesser TSS treatment technologies. Organic chemical facilities are substantially more likely to employ a greater treatment technology (28 of 50 facilities). However, this difference proves statistically insignificant (p=0.451). Second, we consider BOD treatment technologies. Initially, we assess the presence versus the absence of any BOD treatment technology. Other chemical facilities are most likely to operate a BOD treatment technology (94 percent of facilities). Moreover, organic chemical facilities are much more likely than inorganic chemical facilities to operate a BOD treatment technology: 92 percent of facilities versus 54 percent of facilities, which proves statistically significant (p=0.009). We also assess the distinction between greater and lesser BOD treatment technologies. Organic chemical facilities are much more likely than inorganic chemical facilities to employ a greater BOD treatment technology (40 of 51 facilities); again, this difference proves statistically significant ($p = 0.000$).

Next, we assess how upgrades to treatment operations vary across the three broad sectors. Inorganic chemical facilities are much more likely than organic chemical facilities to upgrade (23 of 27 facilities versus 31 of 50 facilities). However, the prevalence of upgrades does not differ between the two primary broad sectors ($p = 0.303$). Within the set of upgrades, we next assess upgrades designed to reduce wastewater discharges. Consistent with the full set of upgrades, inorganic chemical facilities are more likely than organic chemical facilities to upgrade with the intent of reducing discharges, yet the difference proves statistically insignificant ($p = 0.578$).

The extent of a facility's monitoring program may vary across the three broad sectors. Organic chemical facilities are more likely than inorganic chemical facilities to employ the most extensive monitoring program (30 of 51 facilities versus 12 of 28 facilities). However, this difference between the two primary broad sectors is not statistically significant ($p = 0.746$).

The number of audits conducted per year may also differ across the broad sectors. The mean number of audits differs dramatically across the three broad sectors. The two-sample test indicates that organic chemical manufacturing facilities audit more frequently than do inorganic chemical manufacturing facilities (7.2 versus 2.7, $p = 0.040$). Similar to the quantity of audits, we explore whether the quality of audits, as measured by audit team composition, varies across the three broad sectors. Inorganic chemical facilities are substantially more likely than organic chemical facilities to utilize an audit team with the richest composition (21 of 27 facilities versus 27 of 49 facilities). However, the difference between the two primary broad sectors is not significant ($p = 0.332$). As the second measure of audit quality, we assess whether the audit classification protocol—findings ranked by degree of vulnerability for noncompliance versus less extensive protocols—varies across the three broad sectors. Organic chemical facilities are more likely than inorganic chemical facilities to employ the most extensive protocol (13 of 50 facilities versus 5 of 28 facilities). However, the difference proves statistically insignificant ($p = 0.999$).

We next assess the use of external consulting in three forms: whether or not a facility employed an external consultant, the annual amount of money spent on consulting in absolute terms, and the annual amount of money spent in terms relative to the facility's size (as measured by the overall number of employed workers averaged over the relevant three-year period). The presence of consultant use is greatest in the other-chemicals sector (82 percent) and lower in the organic-chemicals sector than in the inorganic-chemicals sector (77 percent versus 78 percent). However, the trivial difference between the two primary sectors proves statistically insignificant (p = 1.000). When considering the amount of money spent on external consultants, sector-specific mean and median values reveal that inorganic chemical facilities spend more money than do organic chemical facilities (respectively, \$46,708 versus \$28,646 and \$20,000 versus \$10,000). However, the two-sample test indicates that the absolute amount of money spent on external consultants does not significantly differ between these two primary broad sectors (p = 0.262). Sector-specific mean and median values reveal a similar comparison when considering external consulting expenditures relative to facility size. However, again this difference proves statistically insignificant (p = 0.150).

Facilities operating in different sectors may seek compliance assistance from their regulators with different proclivities. Inorganic chemical manufacturing facilities are more likely to seek this assistance (39 percent) than both organic chemical manufacturing facilities (22 percent) and other chemical manufacturing facilities (12 percent). However, the difference between the two primary broad sectors proves statistically insignificant (p = 0.657).

Facilities manufacturing other chemicals appear best at providing compliance-assistance materials that meet their facilities' needs. Organic chemical manufacturing facilities appear better at delivering compliance-assistance materials that meet their facilities' needs than do inorganic chemical manufacturing facilities (16 of 46 facilities versus 7 of 27 facilities). However, the difference between the two primary broad sectors proves statistically insignificant (p = 0.999).

Management's concern over noncompliance with wastewater discharge limits may differ across the three broad sectors. Management at inorganic chemical manufacturing facilities is more likely to place a greater concern on noncompliance (93 percent of facilities) than management at organic chemical manufacturing facilities (80 percent of facilities). However, this difference proves statistically insignificant (p = 0.942).

Management communicates its concerns internally and externally. Inorganic chemical manufacturing facilities are more likely to utilize an above-midpoint number of means for communicating their environmental goals internally than are organic chemical manufacturing facilities (16 of 28 facilities versus 25 of 52 facilities). However, this difference proves statistically insignificant (p = 0.998). We also explore the extent of internal communications by assessing whether the number of internal communication means differs across the broad sectors. The two-sample test indicates that the number of internal communication means does not differ between the two primary broad sectors (p = 0.185).

Regarding external communications, inorganic chemical manufacturing facilities and organic chemical manufacturing facilities appear similarly likely to utilize an above-midpoint number of means for communicating their environmental goals externally (6 of 28 facilities versus 12 of 52 facilities). The two-sample test confirms this point since the difference between these two primary broad sectors proves statistically insignificant (p = 1.000). We also explore the extent of external communications by assessing whether the number of external communication means differs across the broad sectors. The two-sample test identifies no significant difference between the two primary broad sectors (p = 0.337).

As the final component of our assessment of environmental management differences across broad sectors, we explore the prevalence of ISO 14001 compliance. Organic-chemicals facilities are more likely to be ISO 14001 compliant (17 percent) than are either inorganic-chemicals facilities (7 percent) or other-chemicals facilities (7 percent). The difference between the two primary sectors proves statistically insignificant (p = 0.996).

5A.1.3. Subsamples Based on Facility Size

In this section, we assess differences in environmental management practices between smaller facilities and larger facilities. The full tabular display is provided on the Web site of Stanford University Press. For this assessment, we split the sample into two subsamples based on the sample median facility size, as measured by the overall number of employees working at an individual facility. We identify the median facility size separately for each data set as distinguished by the relevant time frame: (1) number of workers employed in a specific year when management data are gathered on an annual basis between 1999 and 2001, (2) number of workers employed in 2001 when management data reflect the twelve-month period preceding survey completion, and (3) average number of workers employed between 1999 and 2001 or between 2000 and 2001, depending on the date of survey completion, when management data reflect the three-year period preceding survey completion. The median facility sizes for these three data sets are, respectively, 267 employees, 231.5 employees, and 260 employees.[2]

Based on this division, we test for differences between smaller facilities (that is, below median size) and larger facilities (that is, above median size). When measuring a particular management practice as falling within a set of two categories—for example, the presence or absence of ISO 14001 compliance—we use a two-sample test. When measuring a particular management practice across a spectrum of values, we employ two two-sample tests: Wilcoxon two-sample test and median two-sample test.[3] In the tabular display, we report only the former test results since the latter test results lead to identical conclusions (except environmental employ-ees' education, where the Wilcoxon test and median test p-values equal 0.060 and 0.137, respectively.) In addition, we calculate the correlation between the extent of a management practice and facility size.

We begin our assessment by exploring the allocation of labor devoted to environmental management. Consider first the number of employees devoted to environmental management in general. Smaller facilities em-ploy fewer environmental employees in an absolute sense but more envi-ronmental employees relative to the overall number of employees. Both differences prove statistically significant. We also explore the correlation between the absolute or relative number of environmental employees and facility size (as measured by the number of overall facility employ-ees). The absolute number of environmental employees is positively cor-related with facility size ($\rho = 0.419$) at a highly significant level ($p = 0.0001$, $N = 276$). The relative number of environmental employees is nega-tively correlated with facility size ($\rho = -0.237$) at a highly significant level ($p = 0.0001$, $N = 273$). Thus, as facilities employ more workers, the num-ber of environmental employees grows but does not grow as quickly as the number of overall employees grows. These two results indicate that facilities do not maintain a steady ratio of environmental employees to overall employees as they grow. Instead, facilities spread the number of environmental employees over an expanding worker base.

We also explore the allocation of labor devoted exclusively to waste-water management. The number of wastewater employees is greater in smaller facilities on both an absolute and a relative sense. Both differences are statistically significant. We also assess the correlation between the ab-solute or relative number of wastewater employees and facility size. The absolute number of wastewater employees and the number of overall employees are positively correlated ($\rho = 0.018$) but at a statistically insig-nificant level ($p = 0.86$); in contrast, the relative number of wastewater employees and the number of overall employees are negatively correlated ($\rho = -0.339$) at a significant level ($p = 0.001$). These two results indicate that, as facilities add more workers, they do not increase the number of wastewater employees, allowing the ratio of wastewater employees to overall employees to decline. The two-sample test results indicate that facilities actually decrease the absolute number of wastewater employees as they grow. While the two assessments do not agree on this point, the main thrust remains the same: larger facilities proportionally allocated fewer employees to wastewater management.

In addition to the quantity of labor devoted to environmental man-agement, we assess the quality of labor. Consider first the education of the environmental employees, as measured by the percentage of environmental employees with at least a college degree. Larger facilities employ a significantly greater prevalence of college-educated workers. We also assess the correlation between the degree of college-educated environmental employees and facility size. While larger facilities appear to employ a greater percentage of college-educated environmental employees ($\rho = 0.143$), this relationship is statistically insignificant ($p = 0.18$). Collectively, these results indicate that the distinction between notably larger facilities and notably smaller facilities proves more important than any overall connection between facility size and the education of environmental workers.

As another measure of labor quality, we explore the extent of work experience held by environmental employees and its relationship to facility size. Consider first the distinction between smaller facilities (below median) and larger facilities (above median). Smaller facilities hire environmental employees with more work experience, but this difference is statistically insignificant. Alternatively, we explore the correlation between facility size and environmental employees' level of work experience, which is negative ($\rho = 0.143$) but statistically insignificant ($p = 0.17$). We also explore labor quality by assessing the quality and quantity of training provided by facilities. Initially, we assess whether a facility provides (1) both internal and external training versus (2) internal training only, external training only, or no training. Smaller facilities are slightly more likely to provide both internal and external training than are larger facilities, yet this slight difference is statistically insignificant.

We also explore the quantity of training, as measured by both the number of training days and the extent of attendance at the provided training. Smaller facilities provide more days of training, at least based on the average number of days; based on the median number of days, smaller and larger facilities provide the same number of training days. The two-sample test result reveals that any difference between smaller facilities and larger facilities is statistically insignificant. We also assess the relationship between the number of training days and facility size by calculating the correlation between these two factors. Larger facilities provide fewer days of training ($\rho = -0.098$) but statistically insignificantly so ($p = 0.36$). Thus, the two methods for exploring the relationship between the days of training and facility size generate identical conclusions.

As the second measure of training quantity, we explore the relationship between attendance at any provided training and facility size. A greater percentage of environmental employees working at smaller facilities attend any provided training than environmental employees working at larger facilities, based on the average percentage of attendance. Based on the median values, environmental employees working at smaller and larger facilities both attend all training (100 percent). Despite the identical median values, the two-sample test result indicates that smaller facilities effectively induce a greater percentage of their environmental employees to attend training sessions, though the level of statistical significance is only marginal. We also assess the relationship between training attendance and facility size by calculating the correlation between these two factors. As a facility gets bigger, attendance declines ($\rho = -0.172$), which is consistent with the two-sample test result, though again the level of statistical significance is only marginal ($p = 0.10$).

In addition to labor, facilities devote capital equipment to wastewater management. First, we consider TSS treatment technologies. Initially, we assess the presence versus the absence of any TSS treatment technology. Larger facilities are more likely to employ TSS treatment of any kind; however, the difference proves statistically insignificant. We also assess the distinction between greater and lesser TSS treatment technologies. Again, larger facilities are more likely to employ greater treatment, yet the difference proves statistically significant. Second, we consider BOD treatment technologies. Initially, we assess the presence versus the absence of any BOD treatment technology. Larger facilities are much more likely to employ BOD treatment, as with TSS treatment, and the difference is not statistically significant, as with TSS treatment. We also assess the distinction between greater and lesser BOD treatment technologies. Similar to any BOD treatment, larger facilities are more likely to employ greater BOD treatment. Again, the difference proves statistically insignificant.

Over the three-year period preceding survey completion, some facilities upgraded their treatment equipment. A greater proportion of larger facilities upgraded their treatment equipment, yet this difference is statistically insignificant. Of the noted upgrades, some were designed to reduce wastewater discharges. Similar to any type of upgrade, a greater proportion of larger facilities upgraded their treatment equipment with the intention of reducing wastewater discharges. However, the difference between smaller and larger facilities is less substantial when the upgrades are designed to reduce discharges. The difference seems minor and proves statistically insignificant.

Next we assess the relationship between facilities' efforts to understand their environmental management concerns and facility size. As part of this assessment, we explore the extent of a facility's monitoring program. Larger facilities are more likely to employ the most extensive form of monitoring (which involves both treatment-process monitoring and multiple-point sewer monitoring) than are smaller facilities. However, this difference is statistically insignificant.

We also assess the relationship between facilities' efforts to audit their operations and facility size. We first assess whether the central tendencies or inner quartiles of the audit count, or both, differ between smaller and larger facilities. While the distribution of audits at key percentiles, including the median, does not differ between smaller and larger facilities, the average count of audits is greater for smaller facilities. The two-sample test result indicates that any difference between larger and smaller facilities is statistically insignificant. In addition, we assess whether the annual count of audits correlates with a continuous measure of facility size. Smaller facilities perform fewer audits per year ($\rho = -0.103$) and significantly so, but the level of significance is marginal ($p = 0.102$).

We also explore the quality of self-audits, as measured by the audit team's composition and the extent of the audit classification protocol. When exploring the audit team's composition, we distinguish between teams consisting of both facility employees and visiting corporate employees or visiting consultants and other team compositions, which are regarded as lower in quality. Larger facilities are more likely to utilize a higher-quality audit team, yet this difference is statistically insignificant. Similarly, we explore the relationship between the extent of the audit classification protocol and facility size. We distinguish between classification protocols in which findings are ranked (for example, serially) by the degree of vulnerability for noncompliance and other audit classification protocols, which are regarded as lower in quality. Smaller facilities are more likely to utilize the higher-quality audit classification protocol, yet this difference is statistically insignificant.

Next, we explore the relationship between facilities' efforts to seek insight from others regarding environmental management and facility size. Initially, we explore the relationship between external consultant use and facility size. Larger facilities are more likely to hire an external consultant, yet this difference is statistically insignificant. Moreover, we assess the annual amount of money spent on external consultants, in an absolute sense and relative to facility size. Larger facilities spend more money on external consultants in an absolute sense but less money per facility employee. The first difference is statistically significant, but the second difference is not. In addition, we explore the relationship between external consulting expenditures and facility size by calculating the correlations between these pairs of factors, which reveal that larger facilities spend more absolute money on external consulting ($\rho = 0.170$), though the significance level is only nearly marginal ($p = 0.117$), but larger facilities spend significantly less money per employee ($\rho = -0.236$, $p = 0.028$). These results indicate that, as facilities grow, they increase their spending on external consultants but to an extent that is less than proportional to their size.

We next explore the relationship between facilities' efforts to seek assistance from their regulators as a means for lowering their wastewater discharges and facility size. Smaller facilities are more likely to seek assistance from their regulators than are larger facilities, yet this difference is statistically insignificant. We also explore the success of facilities' efforts to gather information on compliance assistance and facility size. We distinguish between compliance-assistance materials meeting day-to-day needs always and those meeting day-to-day needs less than always. Larger facilities appear more successful than smaller facilities at providing compliance assistance that always meet their day-to-day needs, yet this difference is statistically insignificant.

Next, we explore the relationship between a facility's degree of concern over environmental management and facility size. We distinguish between the highest level of importance ("very important") that management can place on the role of general plant employees in identifying and correcting conditions that may lead to noncompliance and lower levels of importance ("somewhat important" or "important"). Smaller facilities are more likely to place the greatest level of importance on the identified role, yet this difference is statistically insignificant. Related to the level of managerial importance, we explore the relationship between management's efforts to communicate environmental goals and targets to facility employees, as measured by the number of different categories of steps taken, and facility size. Smaller facilities are slightly more likely to employ an above-midpoint count of internal communication methods; this slight difference is statistically insignificant. Similarly, the underlying count of internal communication categories does not differ between smaller and larger facilities. In contrast to internal communication, we next explore the relationship between facility size and the external communication of environmental performance goals and targets to external parties and facility size, as measured by the number of different categories of steps taken. Larger facilities are more likely to employ an above-midpoint count of external communication methods, yet this difference is statistically insignificant. Alternatively, we assess whether the underlying count of external communication categories differs between smaller and larger facilities. Larger facilities take significantly more steps to communicate to external parties.

Finally, we explore the relationship between ISO 14001 compliance and facility size. Larger facilities are slightly more likely to be ISO 14001 compliant; this slight difference is statistically insignificant.

5A.1.4. Subsamples Based on Time:
Years (1999, 2000, 2001)

In this section, we assess differences in environmental behavior over time. The survey gathered data on only two types of environmental management practices that vary over time: the number of environmental employees working at the facility and the number of audits performed each year. We report the number of environmental employees in both absolute terms and terms relative to the overall number of employees working at an individual facility. In the following paragraphs we summarize these two forms of environmental employees for each year between 1999 and 2001. No trend exists in either form. The average absolute quantity moves from 104.7 in 1999 to 98.6 in 2000 and 100.1 in 2001, while the median absolute quantity of 60 remains unchanged. (The year-specific sample sizes are 93, 93, and 94, respectively.) The average relative quantity moves slightly from 0.052 in 1999 to 0.056 in 2000 and 0.055 in 2001, while the median relative quantity also slightly shifts from 0.019 in 1999 to 0.018 in 2000 and 0.020 in 2001. (The year-specific sample size is 91 for each year.) To assess this variation over time, we test whether significant differences exist between the base year of 1999 and the other two years (that is, 1999 versus 2000 and 1999 versus 2001) by employing two-sample tests: Wilcoxon two-sample test and median two-sample test. (We report only the former test results since the latter test results generate identical conclusions.) Based on these two-sample tests, regardless of the form of environmental employees—absolute or relative—the quantity of labor does not differ significantly between 1999 and 2000 or between 1999 and 2001. (Respectively, p-values are 0.721 and 0.972 for the absolute quantities and 0.756 and 0.676 for the relative quantities.)

We also explore the quantity of annual audits over the three-year period between 1999 and 2001. We assess the mean and median values for each year. While the median number of audits, which equals one, does not change over time, the mean number of audits grows over time from 5.15 in 1999 to 5.75 in 2000 and 6.22 in 2001. In order to assess whether the differences over time are statistically significant, we again employ two-sample tests. (We report only the Wilcoxon test results since the median test results generate identical conclusions.) Based on these test results, the number of audits does not differ between 1999 and 2000 or between 1999 and 2001. (Respectively, p-values are 0.991 and 0.461.) These results conform with the lack of variation over time in the median audit value.

5A.2. BEHAVIORAL VARIATION OVER TIME FOR A GIVEN FACILITY

In this section, we explore variation in environmental behavior over time for a given facility. Our survey measures only two environmental management practices over time: number of environmental employees and number of audits per year.

We explore the number of environmental employees in both absolute terms and in relation to the overall number of employees working at a facility. We classify each facility based on the degree of variation over time in the number of environmental employees—in either an absolute sense or as a ratio relative to overall employees. Five facilities (5.2 percent) provide no information on employees. Three facilities (3.1 percent) provide information for two years and employ a positive number of environmental employees in at least one of the two years. Nearly 92 percent (91.8 percent) of the facilities (89 facilities) provide information on environmental employees for all three years. Of these, 52 facilities (54 percent of the sample) employ the same absolute number of environmental employees in each year, yet only 21 facilities (22 percent of the sample) employ the same ratio of environmental employees to overall employees in each year. The remaining facilities vary the number of environmental employees—absolutely or relatively—to some degree over the three-year period. We explore these facilities in more detail. Based on the absolute number, 23 facilities (24 percent of the sample) vary their environmental employees between two consecutive years, but only once, while 14 facilities (14 percent of the sample) vary their environmental employees between two consecutive years twice. Based on the ratio to overall employees, 22 facilities (23 percent of the sample) vary their environmental employees between two consecutive years, but only once, while 46 facilities (47 percent of the sample) vary their environmental employees between two consecutive years twice. Clearly, facilities display more variation in the ratio of environmental employees because they vary their number of overall employees over time. (Of the 97 facilities, nearly half—47 facilities—display a different pattern over time between the absolute and relative number of environmental employees.)

We explore in more detail those facilities that vary their environmental employees over time. Twenty-three facilities vary their environmental employees between two consecutive years once. Three facilities raise their number of environmental employees between 1999 and 2000 and then maintain the higher number in 2001. Nine facilities lower their number between 1999 and 2000 and then maintain the lower number in 2001. Five facilities maintain their number between 1999 and 2000 yet increase the number in 2001. Lastly, six facilities maintain their number between 1999 and 2000 yet decrease it in 2001. In addition, 14 facilities vary their number of environmental employees between two consecutive years twice. Some facilities (3) raise then lower their number. Other facilities (2) lower then raise their number. Just as many facilities (2) increase their number twice over the three-year period. Lastly, seven facilities lower their number twice.

To a greater extent, facilities vary the ratio of environmental to overall employees over time. Twenty-three facilities vary their environmental employee ratios between two consecutive years once. Seven facilities raise their ratio between 1999 and 2000 and then maintain the higher ratio in 2001. Six facilities lower their ratio between 1999 and 2000 and then maintain the lower ratio in 2001. Eight facilities maintain their ratio between 1999 and 2000 yet increase the ratio in 2001. One lone facility maintains its ratio between 1999 and 2000 yet decreases its ratio in 2001. In addition, 46 facilities (47 percent of sample) vary their environmental employee ratio between two consecutive years twice. Some facilities (8) raise then lower their ratio. Just as many facilities lower then raise their ratio. Several facilities (22) increase their ratio twice over the three-year period. Lastly, eight facilities lower their ratio twice. Based on this more refined categorization, again, facilities vary their environmental employees relative to overall employees to a greater extent than their vary the absolute number of environmental employees; of the 97 facilities, more than half (53 facilities) display different patterns over time based on the two measures.

Lastly, we explore the number of audits conducted per year at a given facility. We classify each facility based on the degree of variation in the number of audits over time. Seven facilities (7.2 percent) provide no information on audits. Three facilities (3.1 percent) provide information for only a single year. Two facilities (2.1 percent) provide information for two years; both perform an audit in at least one of the two years. Only 87.7 percent of the facilities provide information on audits for all three years. Of these, two facilities never perform an audit. As the clear majority of facilities, 59 facilities (60.8 percent of the sample) perform the same number of audits in each year. The remaining facilities (24 facilities, 24.7 percent) vary their number of audits to some degree over the three-year period. Of these, 11 facilities (11.3 percent of sample) vary their number of audits between two consecutive years but only once. Two facilities lower their audit count between 1999 and 2000 and then maintain the lower count in 2001. Eight facilities maintain their audit count between 1999 and 2000 yet increase the count in 2001. In addition, 13 facilities (13.4 percent of sample) vary their audit count between two consecutive years twice. Some facilities (3) raise then lower their count. Seven facilities lower then raise their count. Three facilities increase their count twice over the three-year period.

5A.3. ALTERNATIVE TWO-SAMPLE TEST FOR
QUANTITATIVE BEHAVIORAL MEASURES: ASSESSING
THE INFLUENCE OF DISCHARGE LIMITS
ON ENVIRONMENTAL BEHAVIOR

In this section, we describe an alternative two-sample test—the two-sample means t-test—for assessing whether quantitative behavioral measures differ between two subsamples based on the discharge-limit level, which represents part of our hypothesis testing in Chapter 5. This alternative test is valid only if the behavioral measure is normally distributed in each subsample. We assess this condition using four distributional tests: Shapiro-Wilk, Kolmogorov-Smirnov, Cramer–von Mises, and Anderson-Darling. These tests assess whether or not the null hypothesis of a normal distribution can be rejected. We apply the criterion that the null hypothesis is rejected when the p-value associated with a given test lies at or below 0.05 (that is, the null hypothesis is rejected at a significance level of 5 percent). Based on the calculated distributional test statistics, we reject the null hypothesis of normality in at least one of the two subsamples in all but 3 of the 52 relevant cases.

In these 3 exceptional cases, a two-sample means t-test might appear valid. First, when examining the influence of a BOD concentration-equivalent limit on external communications, all of the distributional test p-values lie at or above 0.1043 in the below-median-limit subsample, and all of the distributional test p-values lie at or above 0.0907 in the above-median-limit subsample (three of the four p-values lie at or above 0.1169). The two-sample means t-test result confirms the conclusion drawn from the two primary two-sample test results. (Based on an F-test, we are not able to reject the null hypothesis of equal variances: the F-test statistic equals 1.39 and the p-value equals 0.5784. The equal variance t-test statistic equals −0.52 and the p-value equals 0.6101, which lies above the reported primary two-sample test p-value.)

Second, when examining the influence of a TSS concentration-equivalent limit on external communications, the Shapiro-Wilk test p-value equals 0.1105 in the below-median-limit subsample (the other three distributional test p-values lie at or below 0.028), and all of the distributional test p-values lie above 0.15 in the above-median-limit subsample. If the Shapiro-Wilk distributional test is sufficient for failing to reject the null hypothesis of normality, then a two-sample means t-test is valid. The t-test result confirms the conclusion drawn from the two primary two-sample test results. (Based on an F-test, we are not able to reject the null hypothesis of equal variances: the F-test statistic equals 1.47 and the p-value equals 0.4047. The equal variance t-test statistic equals −1.11 and the p-value equals 0.2732, which is highly similar to the reported primary two-sample test p-value.)

Third, when examining the influence of a BOD concentration-equivalent limit on environmental employee experience, the Shapiro-Wilk test p-value equals 0.0786 in the below-median-limit subsample (the other distributional test p-values lie below 0.05) and all of the distributional test p-values lie above 0.15 in the above-median-limit subsample. If the Shapiro-Wilk distributional test is sufficient for failing to reject the null hypothesis of normality, then a two-sample means t-test is valid. (Based on the reported p-value, the Shapiro-Wilk test rejects the null hypothesis of normality at a 10 percent significance level so the validity of a two-sample means t-test is weaker than in the other two exceptional cases.) The t-test result confirms the conclusion drawn from the two primary two-sample test results. (Based on an F-test, we are not able to reject the null hypothesis of equal variances: the F-test statistic equals 1.28 and the p-value equals 0.6857. The equal-variance t-test statistic equals 0.50 and the p-value equals 0.6230, which lies below the reported primary two-sample test p-value.)

5A.4. SAMPLES BASED ON THREE TIME FRAMES
FOR MEASURING ENVIRONMENTAL BEHAVIOR

In this section, we describe the sample used to examine behavior for each time frame explored. First, we describe the sample used to examine behavior recorded in the calendar years 1999, 2000, and 2001. Of the 97 facilities for which the Environmental Protection Agency's Permit Compliance System (EPA PCS) database provides records on effluent limits and wastewater discharges, all 97 facilities possess records for this three-calendar-year period. For each facility, we consider thirty-six months of data on effluent limits and wastewater discharges. Thus, we consider a sample of 3,492 facility-month observations. Of these observations, a particular facility is active in 92 percent of the months. Of the 97 facilities with EPA PCS records, 5 facilities (5.2 percent) are never active, while 87 facilities (89.7 percent) are always active, during the three-calendar-year subsample period.

Second, we describe the sample used to examine behavior recorded in the twelve-month period preceding the month of survey completion. Of the 97 facilities for which the EPA PCS database provides records on effluent limits and wastewater discharges, all 97 facilities possess records for the twelve-month period preceding the month of survey completion.

For each facility, we consider thirteen months of data on effluent limits and wastewater discharges: twelve months preceding the month of survey completion plus the month of survey completion. Thus, we consider a sample of 1,261 facility-month observations. Of these observations, a particular facility is active in 94.6 percent of the months. Of the 97 facilities with EPA PCS records, 4 facilities (4.1 percent) are never active, while 90 facilities (92.8 percent) are always active, during the thirteen-month subsample period.

Third, we describe the sample used to examine behavior recorded in the three-year period preceding the month of survey completion. Of the 97 facilities for which the EPA PCS database provides records on effluent limits and wastewater discharges, all 97 facilities possess records for the three-year period preceding the month of survey completion. For each facility, we consider thirty-seven months of data on effluent limits and wastewater discharges: thirty-six months preceding the month of survey completion plus the month of survey completion. Thus, we consider a sample of 3,589 facility-month observations. Of these observations, a particular facility is active in 93.1 percent of the months. Of the 97 facilities with EPA PCS records, 4 facilities (4.1 percent) are never active, while 87 facilities (89.7 percent) are always active, during the three-year subsample period.

5A.5. ALTERNATIVE MEANS FOR SUMMARIZING
DISCHARGE LIMIT LEVELS TO MATCH
BEHAVIORAL DATA

In this last section, we assess an alternative approach for matching data on discharge-limit levels to the data on environmental behavior. We summarize the monthly data on discharge limits using the median discharge-limit level, rather than the minimum level, for a particular facility over the relevant period. For consistency, we implement this alternative summary by drawing upon the month-specific discharge-limit median when multiple limit levels apply to a single facility for a given pollutant in a particular measurement form (concentration versus quantity).

In the case of concentration-equivalent limits, the use of medians is expected to generate nearly identical results for BOD limits and highly similar results for TSS limits since BOD concentration-equivalent limits based on monthly-specific minima do not differ over the relevant time periods and TSS concentration-equivalent limits based on monthly-specific minima differ over the relevant time periods to only a very minor degree.

Similar logic applies to quantity limits but to a lesser extent. Even in the case of quantity limits, use of facility-specific and time-period-specific discharge-limit medians for the comparison of behavior between limit-level-defined subsamples generates nearly identical results because the time-period-specific median used to split the samples does not depend meaningfully on whether the facility-specific limit is identified by its minimum or median. For example, consider the behavioral form of audit-team composition, which is measured in the twelve-month period preceding the survey, and the TSS production-adjusted quantity limit. We wish to explore the effect of this limit on audit-team composition. We summarize data on TSS production-adjusted quantity limits for each facility over the twelve months preceding the survey plus the month of survey completion using the facility's minimum limit and the facility's median limit. Thus, we have a set of facility-specific limit minima and a set of facility-specific limit medians. Based on the first set, we evenly split the sample by identifying the sample's fiftieth percentile (that is, sample median). Thus, we distinguish between facilities facing a discharge limit that lies at or below the fiftieth percentile and facilities facing a discharge limit that lies above the fiftieth percentile. Based on the second set, we evenly split the sample by identifying the sample's fiftieth percentile. The difference between these two splits proves inconsequential; in other words, only a small portion of the facilities switch from the lower category to the upper category or vice versa. Specifically, 6.1 percent of the facilities switch. All other comparisons show an even smaller portion of facilities switching.

For all these reasons, we do not report or interpret the results generated by using the facility-specific, period-specific median values. We generate further documentation to support our argument that use of a facility-specific, period-specific minimum and use of a facility-specific, period-specific median generate highly comparable results. We explore the percentage of observations that differ between the two summary measures when assessing the two-sample splits and we calculate the correlation between the two discharge-limit summary measures. (A tabular display is provided on the Web site of the Stanford University Press.) The percentage of observations that differ between the two summary measures lies between 0 percent (3 of 12 cases) and 6.06 percent. The pairwise correlations between the two discharge-limit summary measures mostly lie above 0.93 (9 of 12 cases). Two correlations equal 0.86. One correlation equals 0.73. These rather large correlations reveal that the summary measures are very similar.

Environmental Performance:
Facilities' Discharges and Compliance
with Discharge Limits

This chapter explores the outcomes of facilities' behavioral actions taken to control wastewater discharges in the presence of discharge limits. We measure these outcomes in terms of discharges and compliance and refer to them as "environmental performance." As part of this exploration, this chapter analyzes the effects of discharge limits and environmental behavior on environmental performance. In particular, the chapter tests whether tighter discharge limits lower absolute discharges, while raising discharge ratios, and whether better environmental behavior leads to better environmental performance.

6.1. WASTEWATER DISCHARGES MEASURED IN TWO FORMS: QUANTITIES AND CONCENTRATIONS

As described in Chapter 4, facilities' wastewater discharges are measured in two forms: quantities and concentrations. Quantity discharges are recorded in terms of mass of pollutant per day (for example, kilograms of total suspended solids [TSS] per day). Concentration discharges are recorded in terms of the ratio of pollutant mass per unit of flow volume (for example, milligrams of TSS per liter of wastewater flow).

6.2. DATA SOURCE OF
ENVIRONMENTAL PERFORMANCE

As described in Chapter 2, the Environmental Protections Agency's Permit Compliance System (EPA PCS) database provides all necessary information on wastewater discharges—both quantities and concentration—for all regulated pollutants, including TSS and biological oxygen demand (BOD). Again, this database systematically provides discharge data only for major facilities. For these facilities, the database provides monthly data on discharges. For our analysis, we treat each facility's discharge in a given month as the unit of observation.

6.3. ENVIRONMENTAL PERFORMANCE IN
MULTIPLE FORMS

The PCS database records information on absolute discharges—both quantities and concentrations. By relating these absolute discharges to the relevant permitted discharge levels (that is, discharge limits), we are able to calculate relative discharges or "discharge ratio" (that is, ratio of absolute discharges to permitted discharges). This ratio represents the extent of compliance. As the ratio falls, the extent of compliance rises. Finally, we are able to derive compliance status from the discharge ratio. A regulated facility is compliant with a relevant discharge limit if the discharge ratio is equal to or less than one (that is, absolute discharges are equal to or less than permitted discharges). A regulated facility is noncompliant with a relevant discharge limit if the discharge ratio exceeds one. For our sample of facilities, we describe environmental performance with respect to each form of environmental performance, with a strong focus on the discharge ratio since it is the most comprehensive form and facilitates comparison across facilities and across time for any facility whose discharge limit varies over time.

6.4. EXTENT OF COMPLIANCE AND COMPLIANCE
STATUS: TSS AND BOD DISCHARGE LIMITS

In order to assess compliance, our analysis must consider specific pollutants and the relevant pollutant-specific discharge limits. Initially, we consider any pollutant regulated by an operative discharge limit. Given this scope, we assess the availability of data on measured wastewater discharges. Of the 4,493 facility-month observations possessing a discharge limit for at least one regulated wastewater pollutant, nearly 100 percent of the observations (99.6 percent, 4,473 observations) also provide data on wastewater discharges. This nearly universal reporting of wastewater discharges is quite reassuring and practically eliminates the need to consider strategic nonreporting of discharges. To be fair, this conclusion of nearly full reporting masks the possibility that in a particular month a specific facility faces multiple operative limits yet fails to report wastewater discharges for a subset of the limited pollutants. To address this point, we next examine the study's two key individual pollutants—TSS and BOD.

First, we consider TSS discharges. We address quantity discharges and concentration discharges separately and jointly. Of the 3,269 facility-month observations possessing a quantity-based discharge limit, nearly 100 percent of the observations (99.5 percent, 3,253 observations) also provide data on TSS quantity discharges. Similarly, of the 1,273 facility-month observations possessing a concentration-based discharge limit, nearly 100 percent of the observations (99.5 percent, 1,266 observations) also provide data on TSS concentration discharges. Considered jointly, of the 3,714 facility-month observations possessing either a quantity- or concentration-based discharge limit, nearly 100 percent of the observations (99.5 percent, 3,696 observations) also provide data on the regulated discharges.

Second, we consider BOD discharges. Of the 2,924 facility-month observations possessing a quantity-based discharge limit, nearly 100 percent of the observations (99.7 percent, 2,916 observations) also provide data on BOD quantity discharges. Even better, of the 860 facility-month observations possessing a concentration- based discharge limit, 100 percent of the observations also provide data on BOD concentration discharges. Considered jointly, of the 3,117 facility-month observations possessing either a quantity- or concentration-based discharge limit, nearly 100 percent of the observations (99.7 percent, 3,109 observations) also provide data on the regulated discharges.

Thus, in the specific cases of TSS and BOD, the nearly universal reporting of wastewater discharges is quite reassuring and practically eliminates the need to consider strategic nonreporting of discharges.

6.5. ENVIRONMENTAL PERFORMANCE
FOR CHOSEN SAMPLE

We depict environmental performance in three performance forms—absolute discharges, relative discharges (that is, discharge ratio, extent of compliance), and compliance status—and in two measurement forms—quantities and concentrations. When examining absolute discharges, we analyze each pollutant separately and each measurement form separately. (Thus, we do not attempt to convert quantities into concentration-equivalent terms.) In the case of absolute quantity discharges, we scale each facility's quantity discharges by the facility's production level using the same protocol to scale quantity-based discharge limits by production level (hereafter "production-adjusted absolute quantity discharges"). When examining compliance, we first analyze each combination of pollutant and measurement form separately. Then we combine these elements. We combine the concentration-based discharge ratio and the quantity-based discharge ratio in those cases where we generated a concentration-equivalent discharge limit based on the presence of an operative discharge limit imposed on the facility's wastewater flow (hereafter "concentration-equivalent discharge ratio"). In this combination, the concentration-based discharge ratio takes precedence. We also combine the quantity-based discharge ratio and the concentration-based discharge ratio for each pollutant separately by averaging the two levels when both are present (hereafter "quantity- and concentration-based discharge ratio"). Lastly, we combine the TSS-related quantity- and concentration-based discharge ratio and the BOD-related quantity- and concentration-based discharge ratio by averaging the two levels when both are present (hereafter "TSS and BOD quantity- and concentration-based discharge ratio").

As with discharge limits, we acknowledge that the data on wastewater discharges reported in the PCS database represent information from multiple pipes. Thus, multiple discharge levels may relate to a specific facility in a given month. As with discharge limits, we address the noted multiplicity of discharges by aggregating the data on discharges to the combination of a particular facility and specific month within one year ("triplet"). When calculating absolute discharges and discharge ratios, we employ two aggregation protocols: mean within each triplet and median within each triplet. When calculating discharge ratios, we also identify the maximum within each triplet. When calculating compliance status, we identify only the maximum within each triplet (that is, the worst degree of noncompliance identifies compliance status for a given facility and month). The mean-aggregated absolute discharges and median-aggregated absolute discharges are nearly identical in all cases, with the exception of TSS production-adjusted absolute quantity discharges; in this exceptional case, the two levels are highly similar except at the lower and upper reaches of the two distributions. The mean-aggregated discharge ratios and median-aggregated discharge ratios are nearly identical in all cases. The maximum-aggregated discharge ratios are only slightly greater than the other aggregated ratios. Thus, it seems legitimate to use any one of the aggregated performance levels in general. We focus exclusively on the median-aggregated protocol since it is not sensitive to outliers, unlike the other two aggregation protocols.[1]

For the assessment of wastewater discharges, for each pollutant, we separately explore the distribution of absolute quantity discharges, production-adjusted absolute quantity discharges, absolute concentration discharges, quantity-based discharge ratios, concentration-based discharge ratios, concentration-equivalent discharge ratios, quantity- and concentration-based discharge ratios, quantity-based compliance status, concentration-based compliance status, and quantity- and concentration-based compliance status. For the two pollutants jointly, we consider quantity- and concentration-based discharge ratios and quantity- and concentration-based compliance status. For this exploration, we report the average, median, and key percentiles.

Consider first TSS discharges, as shown in Table 6.1.a. The average facility discharges 757 kg of TSS per day, while the median facility discharges only 28 kg of TSS per day. Clearly, the distribution is skewed toward larger quantities. After adjusting for production, the average facility discharges 6.4 kg of TSS per 1,000 kg of product per day, while the median facility discharges only 0.19 kg of TSS per 1,000 kg of product per day. In the case of concentration-based discharges, the average facility discharges 17.5 mg of TSS per liter of wastewater flow, while the median facility discharges 6.5 mg of TSS per liter. Again, the distribution is skewed toward larger concentration values but not as strongly as quantity discharges. By relating these absolute discharges to permitted discharges, we explore discharge ratios. The average facility is substantially overcompliant with both its quantity-based limit and concentration-based limit by discharging only 27 percent and 30 percent, respectively, of its relevant limits. Considering jointly quantity- and concentration-based discharges and limits, based on the mean discharge ratio of 0.267, the average facility discharges TSS at levels 73 percent below its limits ([1 − 0.267] × 100 = 73.3 percent). The median facility discharges at even lower levels: 19 percent, 20 percent, and 19 percent, respectively, of its quantity-based limits, concentration-based limits, and quantity- and concentration-based limits. The average facility is noncompliant with its TSS quantity-based limit, TSS concentration-based limit, or either limit (jointly considered) in only 1.78 percent, 3.79 percent, and 2.81 percent of the months in the sample period, as shown in Table 6.1.b.

An exploration of BOD discharges reveals a highly similar pattern, as shown in Table 6.1.c. The average facility discharges 93 kg of BOD per day, while the median facility discharges only 18 kg of BOD per day.

TABLE 6.1

Discharges: Summary statistics

Table 6.1.a. TSS discharges

Discharge Measure	N	Mean	25th Percentile	Median	75th Percentile
Absolute Quantity without Adjustment for Facility-Specific Production	3,719	756.907	3.629	28.168	97.069
Absolute Quantity with Adjustment for Facility-Specific Production	3,719	6.355	0.008	0.192	1.209
Absolute Concentration	2,203	17.542	0.000	6.500	16.000
Quantity Discharge Ratio	3,253	0.267	0.091	0.194	0.355
Concentration Discharge Ratio	1,266	0.297	0.060	0.203	0.378
Concentration-Equivalent Discharge Ratio	1,650	0.269	0.063	0.187	0.353
Quantity and Concentration (jointly considered) Discharge Ratio	3,696	0.267	0.082	0.192	0.356

Table 6.1.b. TSS discharges compliance status

Limit Type	Category	Frequency	%
Quantity	Noncompliant	58	1.78
	Compliant	3,195	98.22
Concentration	Noncompliant	48	3.79
	Compliant	1,218	96.21
Quantity and Concentration (jointly considered)	Noncompliant	104	2.81
	Compliant	3,592	97.19

SOURCE: Environmental Protection Agency Permit Compliance System database.

(continued)

Again, the distribution is skewed toward larger quantities. After adjusting for production, the average facility discharges 5 kg of BOD per 1,000 kg of product per day, while the median facility discharges only 0.13 kg of BOD per 1,000 kg of product per day. In the case of concentration-based discharges, the average facility discharges 27.5 mg of BOD per liter of wastewater flow, while the median facility discharges 4.5 mg of BOD per liter. Again, the distribution is skewed toward larger concentration values. The average facility is substantially overcompliant with both its quantity-based limit and concentration-based limit by discharging only 25 percent and 24 percent, respectively, of its relevant limits. Considering jointly quantity- and concentration-based discharges and limits, based on the mean discharge ratio of 0.261, the average facility discharges BOD at levels 74 percent below its limits ([1 − 0.261] × 100 = 73.9 percent). The

TABLE 6.1 *(continued)*

Table 6.1.c. BOD discharges

Discharge Measure	N	Mean	25th percentile	Median	75th percentile
Absolute Quantity Without Adjustment for Facility-Specific Production	3,047	95.123	3.538	18.597	60.328
Absolute Quantity with Adjustment for Facility-Specific Production	3,047	4.973	0.012	0.134	1.160
Absolute Concentration	1,381	27.536	2.000	4.500	12.000
Quantity Discharge Ratio	2,916	0.362	0.064	0.148	0.293
Concentration Discharge Ratio	860	0.239	0.057	0.186	0.348
Concentration-Equivalent Discharge Ratio	1,181	0.213	0.064	0.141	0.300
Quantity and Concentration (jointly considered) Discharge Ratio	3,109	0.364	0.068	0.163	0.314

Table 6.1.d. BOD discharges compliance status

Limit Type	Category	Frequency	%
Quantity	Noncompliant	89	3.05
	Compliant	2,827	96.95
Concentration	Noncompliant	4	0.47
	Compliant	856	99.53
Quantity and Concentration (jointly considered)	Noncompliant	92	2.96
	Compliant	3,017	97.04

Table 6.1.e. TSS and BOD discharges, quantity and concentration (jointly considered): Discharge ratio

	N	Mean	25th percentile	Median	75th percentile
Discharge Ratio	3,864	0.358	0.088	0.196	0.342

Table 6.1.f. TSS and BOD discharges, quantity and concentration, (jointly considered): Compliance status

Compliance Status	Frequency	%
Noncompliant	119	3.08
Compliant	3,745	96.92

SOURCE: Environmental Protection Agency Permit Compliance System database.

median facility discharges at even lower levels: 15 percent, 19 percent, and 16 percent, respectively, of its quantity-based limits, concentration-based limits, and quantity- and concentration-based limits. The average facility is noncompliant with its BOD quantity-based limit, concentration-based limit, or either limit in only 3.05 percent, 0.47 percent, and 2.96 percent of the months in the sample period, as shown in Table 6.1.d.

As the most comprehensive measure of environmental performance, the TSS and BOD quantity- and concentration-based discharge ratio equals 0.28 on average, with a median of 0.20, as shown in Table 6.1.e. Thus, the median facility is 80 percent overcompliant with TSS and BOD limits in general. The average facility is noncompliant in only 3.08 percent of the months in the sample period, as shown in Table 6.1.f.

This exploration examines environmental performance for the entire sample, without any reference to other related factors, such as facility size. In the appendix to this chapter, we depict the same data with reference to certain related factors: facility location, broad manufacturing subsector, facility size, and time. In the appendix, we also assess variation in environmental performance across time for a given facility. For this depiction and assessment, we explore only the most comprehensive measure of environmental performance for each pollutant: quantity- and concentration-based discharge ratio. As a prelude, Figure 6.1 graphs the median discharge ratio for TSS and BOD over the sample period on a monthly basis. This figure reveals no dramatic trend for either pollutant. While focusing on TSS and BOD discharges jointly, Figure 6.2 graphs the median discharge ratio

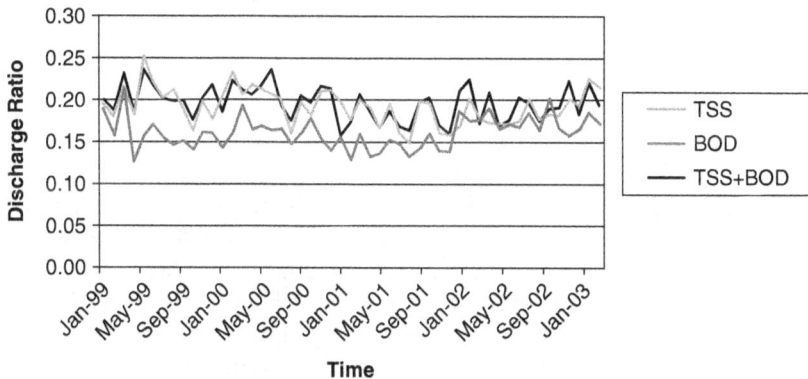

FIGURE 6.1

Discharge ratio of median facility

SOURCE: Environmental Protection Agency Permit Compliance System database.

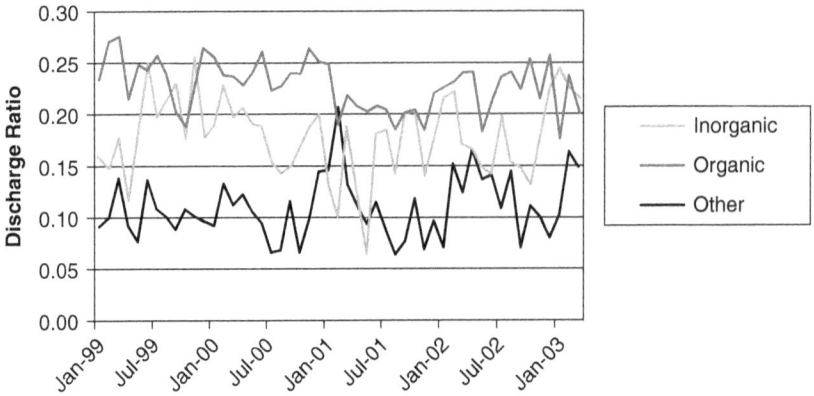

FIGURE 6.2

Sector-specific median discharge ratio

SOURCE: Environmental Protection Agency Permit Compliance System database.

separately for each chemical manufacturing subsector—organic, inorganic, and other—over the sample period. The discharge ratio for organic chemical facilities generally lies above the other two subsectors, while the discharge ratio for other chemical facilities generally lies below the other two subsectors. None of the sectors displays a dramatic trend over the sample period.

6.6. THE EFFECT OF DISCHARGE LIMITS
ON ENVIRONMENTAL PERFORMANCE

6.6.1. How Do Discharge Limits Affect
Environmental Performance?

Chapter 4 describes the discharge limits imposed on regulated facilities discharging wastewater. The preceding sections of this chapter describe the levels of pollution discharged by the regulated facilities when facing these imposed limits. In this section, we probe the obvious connection between these two elements: how do imposed discharge limits affect environmental performance as measured by both absolute discharges and absolute discharges relative to permitted discharges (that is, discharge ratio)?

This research question relates to two forms of environmental performance: absolute discharges and discharge ratios. However, the interpretation of the answer differs between the two forms. Thus, we attempt to answer the question by separately analyzing each form. Based on our analysis of absolute discharges, if the analytical results reveal that absolute discharges fall as limits tighten (in reverse, absolute discharges rise as limits loosen), then discharge limits appear effective at prompting better environmental performance. Clearly, this conclusion possesses very strong policy implications. If environmental agencies wish to reduce absolute discharges, tighter limits would appear an effective tool. Based on our analysis of discharge ratios, if the analytical results reveal that discharge ratios rise as limits tighten (in reverse, discharge ratios fall as limits loosen), then discharge limits appear not only effective but binding. By *binding*, we mean that facilities are constrained by the limits. Consistent with this constraint, facilities are expected to struggle more in order to comply with tighter discharge limits than they struggle to comply with looser discharge limits. As evidence of this greater struggle, facilities choose to let their discharge ratio rise (that is, the extent of compliance slip) as limits tighten. Put differently, as limits tighten, facilities face higher control costs—on both a total basis and a marginal basis, which prompts them to balance control costs and the costs of noncompliance at a higher discharge ratio (that is, lower extent of compliance).

6.6.2. Analytical Approach

In order to test the two hypotheses associated with the effects of limits on environmental performance, we examine the effects of discharge limits on two measures of environmental performance—absolute discharges and discharge ratios—that are specific to the pollutant and specific to the measurement form: quantity and concentration. We employ two analytical strategies. As one strategy, we assess whether the distribution of the environmental performance measure differs between two subsamples: facilities facing below-median discharge limits and facilities facing above-median discharge limits. We first identify the distribution in each subsample and then compare the distributions. According to the first hypothesis, we expect that facilities facing tighter limits (below-median limits) generate a lower distribution of absolute discharges than do facilities facing looser limits (above-median limits). By lower distribution, we mean a distribution located at lower values of absolute discharges. According to the second hypothesis, we expect that facilities facing tighter limits (below-median limits) generate a higher distribution of discharge ratios than do facilities facing looser limits (above-median limits). By higher distribution, we mean a distribution located at higher values of the discharge ratio. As the second analytical strategy, we assess the correlation between an environmental performance measure and a discharge-limit level. According to the first hypothesis, we expect a positive correlation between discharge limits and absolute discharges. According to the second hypothesis, we expect a negative correlation between discharge limits and discharge ratios.

Readers less interested in the details of our statistical analysis than in the results of our statistical analysis may wish to skip over the following two paragraphs.

As the final component of the first analytical strategy, we employ two-sample tests—the Wilcoxon two-sample test and the median two-sample test—in order to assess whether any identified difference in the subsample-specific distributions is statistically significant. (We report only the former test results because the latter test results generate identical conclusions.)[2] These two-sample tests employ scoring methods for assessing whether the distribution of one subsample is located at higher levels of absolute or relative discharges (that is, discharge ratio) than the distribution of the other subsample. Any significant difference between the two distributions provides evidence for or against the two hypotheses. We deem a difference between two distributions statistically significant when our analysis is able to reject the null hypothesis of zero difference at a significance level of 10 percent or less, as revealed by a p-value of 0.10 or less. We deem a difference between two distributions statistically insignificant when our analysis is not able to reject the null hypothesis of zero difference at a significance level of 10 percent or less, as revealed by a p-value greater than 0.10.[3]

For our second analytical strategy, we assess the correlation between the environmental performance measure and the discharge-limit level by calculating the Pearson correlation coefficients associated with the relevant pairwise comparisons of individual environmental performance measures and discharge limits. Each correlation captures whether a discharge-limit level and performance measure vary together in a systematic fashion. For our exploration, we first assess the signs of these correlations in order to explore the types of relationships between each performance measure and discharge limit. A positive correlation indicates that the level of absolute or relative discharges and the level of the discharge limit generally vary together in the same direction. A negative correlation indicates that the level of absolute or relative discharges and level of the discharge limit generally vary together in opposite directions. We also assess the magnitude of these correlations in order to explore the strength of the relationships between environmental performance and discharge limits. In addition, we assess the statistical significance of these correlations, as captured by the p-values associated with the correlations, in order to test whether the calculated magnitudes are distinguishable from zero, which implies no relationship between an individual environmental performance measure and an individual discharge-limit level. The calculated level of statistical significance, reflected in the associated p-value, stems from a comparison of the correlation magnitude and the underlying variance of the two factors. We deem a correlation statistically significant when our analysis is able to reject the null hypothesis of a zero correlation at a significance level of 10 percent or less, as revealed by a p-value of 0.10 or less. We deem a correlation statistically insignificant when our analysis is not able to reject the null hypothesis of a zero correlation at a significance level of 10 percent or less, as revealed by a p-value greater than 0.10.[4]

In Chapter 4, we describe and interpret four forms of discharge limit: quantity, production-adjusted quantity, concentration, and concentration-equivalent. As argued previously, we believe that the production-adjusted quantity limit dominates the quantity limit, so we ignore the effect of the latter limit form on environmental performance.

6.6.3. Analytical Results and Interpretation

We consider first the effects of discharge limits on TSS-related environmental performance. As shown in Table 6.2.a, facilities facing tighter TSS

TABLE 6.2

Hypothesis testing: Effects of discharge limits on environmental performance

6.2.a. TSS discharges

Limit and Discharge Form	Limit Subsample[a]	N	Median	Sign of Difference in Medians[b]	Test Statistic (p-value)[c]
Production-Adjusted Absolute Quantity	Below median	1,627	0.043	+	36.900
	Above median	1,626	1.409		(0.001)
Absolute Concentration	Below median	854	5.000	+	10.360
	Above median	412	12.000		(0.001)
Quantity Discharge Ratio	Below median	1,627	0.214	–	−5.756
	Above median	1,626	0.175		(0.001)
Concentration Discharge Ratio	Below median	854	0.167	+	8.664
	Above median	412	0.301		(0.001)
Concentration-Equivalent Discharge Ratio	Below median	1,049	0.175	+	5.363
	Above median	601	0.200		(0.001)

[a] Limit categories: below-median limit, above-median limit.
[b] The difference in median values = (median of above-median subsample) – (median of below-median subsample).
[c] Test statistic applies to the comparison between the below-median-limit subsample and above-median-limit subsample; reports the Wilcoxon two-sample test in the form of a normal approximation. The p-value applies to a one-sided test.

(continued)

TABLE 6.2 *(continued)*

6.2.b. BOD discharges

Limit and Discharge Form	Limit Subsample[a]	N	Median	Sign of Difference in Medians[b]	Test Statistic (p-value)[c]
Production-Adjusted	Below median	1,450	0.023	+	36.731
Absolute Quantity	Above median	1,449	1.168		(0.001)
Absolute Concentration	Below median	509	3.000	+	6.994
	Above median	351	5.000		(0.001)
Quantity Compliance	Below median	1,450	0.149	−	−0.192
Ratio	Above median	1,449	0.142		(0.424)
Concentration Compliance	Below median	509	0.225	−	−4.875
Ratio	Above median	351	0.114		(0.001)
Concentration-Equivalent	Below median	600	0.190	−	−7.350
Compliance Ratio	Above median	581	0.097		(0.001)

[a] Limit categories: below-median limit, above-median limit.

[b] The difference in median values = (median of above-median subsample) - (median of below-median subsample).

[c] Test statistic applies to the comparison between the below-median-limit subsample and above-median-limit subsample; it reports the Wilcoxon two-sample test in the form of a normal approximation. The p-value applies to a one-sided test.

quantity-based limits (below-median limits) discharge significantly lower absolute quantities of TSS than do facilities facing looser TSS quantity-based limits (above-median limits). Figure 6.3 (top) displays this effect even more dramatically. As further evidence, based on the correlation, as TSS quantity-based limits tighten, facilities' TSS absolute quantity discharges significantly fall ($\rho = 0.614$, $p < 0.0001$); that is, as quantity-based limits rise, facilities' TSS absolute quantity discharges significantly increase. A similar conclusion applies to TSS absolute concentration discharges, but the evidence is weaker. As shown in Table 6.2.a, facilities facing tighter TSS concentration-based limits (below-median limits) discharge significantly lower absolute concentrations of TSS than do facilities facing looser TSS concentration-based limits (above-median limits). Based on the correlation, as TSS concentration-based limits tighten, facilities' TSS absolute concentration discharges fall ($\rho = 0.017$). However, this relationship is not statistically significant ($p = 0.272$). In general, these results indicate that TSS discharge limits are effective.

The second testable hypothesis relates to the effect of discharge limits on discharge ratios. As shown in Table 6.2.a, facilities facing tighter TSS quantity-based limits (below median) generate significantly higher quantity-based discharge ratios than do facilities facing looser TSS quantity-based

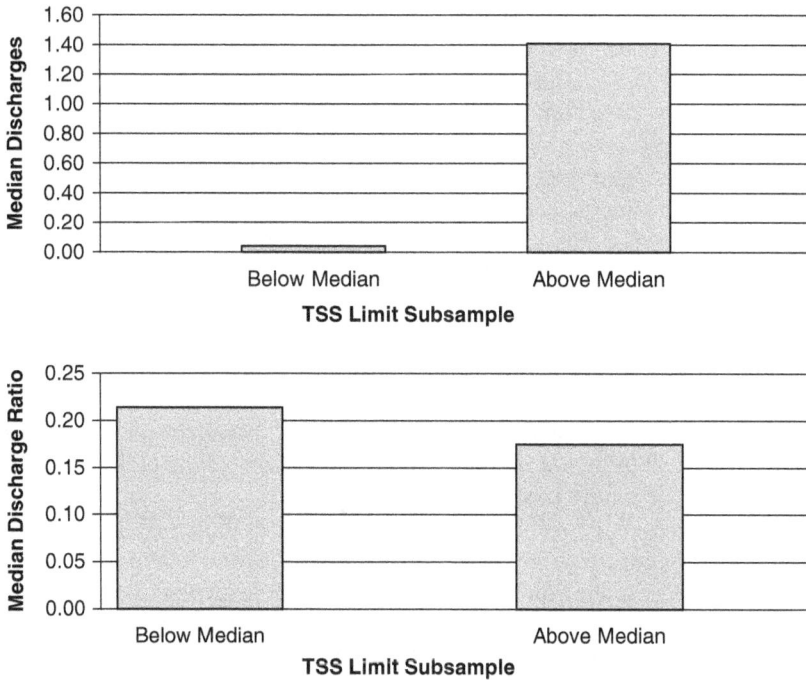

FIGURE 6.3

Effect of TSS limits on TSS quantity-based discharges. Top: Production-adjusted TSS absolute discharges. Bottom: TSS quantity-based discharge ratio.

NOTE: Production-adjusted absolute quantity discharges measured in kilograms of TSS per 1,000 kilograms of product per day.

limits (above median). In other words, facilities facing tighter TSS quantity-based limits comply less fully with these limits than do facilities facing looser TSS quantity-based limits. Figure 6.3 (bottom) displays this effect also. As further evidence, based on the correlation, as TSS quantity-based limits tighten, facilities' TSS-related quantity-based discharge ratios significantly fall ($\rho = -0.080$, $p < 0.0001$); that is, as quantity-based limits rise, facilities' TSS-related quantity-based discharge ratios significantly decrease.[5]

Contrary to our expectations, the opposite conclusion applies to TSS concentration-related discharge ratios. As shown in Table 6.2.a, facilities facing tighter TSS concentration-based limits (below median) generate a significantly lower concentration-based discharge ratio than do facilities facing looser TSS concentration-based limits (above median). In other words, facilities facing tighter TSS concentration-based limits comply more fully with these limits than do facilities facing looser TSS concentration-based limits. As further evidence, based on the correlation, as TSS concentration-based limits tighten, facilities' TSS-related concentration-based discharge ratios significantly rise ($\rho = 0.087$, $p = 0.001$); that is, as concentration-based limits rise, facilities' concentration-based discharge ratios significantly increase. Results of the relationship between concentration-equivalent limits and compliance ratios are highly similar, as shown in Table 6.2.a (relevant correlation equals 0.043, $p = 0.041$). These results indicate that tighter concentration-based TSS discharge limits appear to induce improved compliance, perhaps by prompting facilities to discover and implement inventive means of controlling TSS discharges. This conclusion is consistent with the Porter hypothesis, which claims that tighter performance-based standards induce facilities to locate cheaper means of controlling pollution, not simply better means of controlling pollution (Porter and van der Linde, 1995).

These two sets of results indicate that quantity-based TSS discharge limits are binding yet concentration-based TSS discharge limits appear to induce cheaper pollution control. As one possible explanation for this difference, perhaps concentration-based limits grant facilities more flexibility to innovate than do quantity-based limits. This greater flexibility is not surprising since concentration-based limits less strongly constrain manipulations of production levels.

We consider next the effects of discharge limits on BOD-related environmental performance. As shown in Table 6.2.b, facilities facing tighter BOD quantity-based limits (below median) discharge significantly lower absolute quantities of BOD than do facilities facing looser BOD quantity-based limits (above median). Moreover, based on the correlation, as BOD quantity-based limits tighten, facilities' BOD absolute quantity discharges significantly fall ($\rho = 0.655$, $p < 0.0001$); that is, as quantity-based limits rise, facilities' BOD absolute quantity discharges significantly increase. An identical conclusion applies to BOD absolute concentration discharges. As shown in Table 6.2.b, facilities facing tighter BOD concentration-based limits (below median) discharge significantly lower absolute concentrations of BOD than do facilities facing looser concentration-based limits (above median). Also, based on the correlation, as BOD concentration-based limits tighten, facilities' BOD absolute concentration discharges significantly fall ($\rho = 0.232$, $p < 0.0001$). The magnitude of this correlation coefficient is much greater than the correlation coefficient relating TSS-related concentration-based limits and discharges. These results strongly indicate that BOD discharge limits are effective.

The second testable hypothesis relates to the effect of discharge limits on discharge ratios. As shown in Table 6.2.b, facilities facing tighter BOD quantity-based limits (below median) generate a higher quantity-based discharge ratio than do facilities facing looser BOD quantity-based limits (above median). In other words, facilities facing tighter BOD quantity-based limits comply less fully with these limits than facilities facing looser BOD quantity-based limits. However, the difference is not statistically significant. As stronger evidence, based on the correlation, as BOD quantity-based limits tighten, facilities' BOD-related quantity-based discharge ratios significantly rise ($\rho = -0.033$, $p = 0.039$); that is, as quantity-based limits rise, facilities' quantity-based discharge ratios significantly decrease.

A similar yet stronger conclusion applies to BOD concentration-related discharge ratios. As shown in Table 6.2.b, facilities facing tighter BOD concentration-based limits (below median) generate a significantly higher concentration-based discharge ratio than do facilities facing looser BOD concentration-based limits (above median). In other words, facilities facing tighter BOD concentration-based limits comply less fully with these limits than do facilities facing looser BOD concentration-based limits. As similar evidence, based on the correlation, as BOD concentration-based limits tighten, facilities' BOD-related concentration-based discharge ratios significantly rise ($\rho = -0.259$, $p < 0.0001$); that is, as concentration-based limits rise, facilities' concentration-based discharge ratios significantly decrease. Results of the relationship between concentration-equivalent limits and compliance ratios are highly similar, as shown in Table 6.2.b (correlation of $\rho = 0.206$, $p < 0.0001$).

These results indicate that BOD discharge limits, especially concentration-based limits, are binding.

Overall, these results indicate that both TSS and BOD discharge limits are effective, but only BOD discharge limits are definitively binding since TSS concentration-based limits appear to induce facilities to discover cheaper means of controlling TSS discharges.

6.7. THE EFFECT OF ENVIRONMENTAL BEHAVIOR ON ENVIRONMENTAL PERFORMANCE

6.7.1. How Does Environmental Behavior Affect Environmental Performance?

Chapter 5 describes the environmental behavior shown by facilities as they expend effort to comply with their discharge limits. The preceding sections of this chapter describe the levels of pollution discharged by the regulated facilities as a result of these efforts when facing the imposed discharge limits. In this section, we probe the obvious connection between environmental behavior and performance: how does environmental behavior affect environmental performance as measured by absolute discharges relative to permitted discharges (that is, discharge ratio)?

Clearly, one hopes that better environmental behavior leads to better performance; otherwise, regulated facilities would need to question the need to expend efforts at improving environmental behavior. The effect of behavior on performance also has policy implications. If better environmental behavior leads to better environmental performance, then it might be useful for environmental protection agencies to assess environmental behavior even though discharge limits are performance based. The expectation that better behavior leads to better performance represents a key hypothesis to test.

6.7.2. Analytical Approach

In order to test this hypothesis, we examine the effects of various measures of environmental behavior on the most comprehensive measure of environmental performance: TSS and BOD quantity- and concentration-based discharge ratio. We explore each environmental behavioral measure in turn. (Given the demonstrated variation in discharge limits across facilities and time, analysis of absolute discharges seems dominated by analysis of discharge ratios, which relate absolute discharges to facility-period-specific discharge limits; moreover, given the large number of behavioral measures, analysis of pollutant-specific or measurement-specific measures of discharge ratios, for example, TSS-related quantity-based discharge ratio, seems a bit unwieldy.) We employ two analytical strategies. As one strategy, we assess whether the distribution of the environmental performance measure differs between two subsamples. In the case of qualitative measures of environmental behavior, we identify these two subsamples: facilities employing lesser environmental behavior and facilities employing greater environmental behavior (as described in Chapter 5). In the case of quantitative measures of environmental behavior, we identify lesser environmental behavior as below-median behavior and greater environmental behavior as above-median behavior; the relevant median values are identified in Chapter 5. Given these two subsamples, we identify the distribution of environmental performance levels in each subsample and then compare the distributions. Based on our hypothesis, we expect that facilities expending greater environmental behavior generate a lower distribution of discharge ratios than do facilities employing lesser environmental behavior.

As the second analytical strategy, we assess the correlation between the level of environmental performance and the extent of environmental behavior in the case of quantitative measures. According to our hypothesis, we expect a negative correlation between these two factors.

Readers less interested in the details of our statistical analysis than in the results of our statistical analysis may wish to skip over the following two paragraphs or even to the beginning of Section 6.7.3, in which we interpret the results of our hypothesis testing.

As the final component of the first analytical strategy, we employ two-sample tests—the Wilcoxon two-sample test and the median two-sample test—in order to assess whether any identified difference in the subsample-specific distributions is statistically significant. (We report only the former test results because the latter test results generate identical conclusions except as noted.)[6] These two-sample tests employ scoring methods for assessing whether the distribution of one subsample is located at higher levels of relative discharges (that is, discharge ratio) than the distribution of the other subsample. Any significant difference between the two distributions provides evidence for or against our hypothesis. We deem a difference between two distributions statistically significant when our analysis is able to reject the null hypothesis of zero difference at a significance level of 10 percent or less, as revealed by a p-value of 0.10 or less. We deem a difference between two distributions statistically insignificant when our analysis is not able to reject the null hypothesis of zero difference at a significance level of 10 percent or less, as revealed by a p-value greater than 0.10.[7]

For our second analytical strategy, we assess the correlation between a quantitative measure of environmental behavior and an environmental performance measure by calculating the Pearson correlation coefficients associated with the relevant pairwise comparisons of individual environmental behavior measures and individual environmental performance measures. Each correlation captures whether an environmental behavior measure and an environmental performance measure vary together in a systematic fashion. For our exploration, we first assess the signs of these correlations in order to explore the type of relationship between each behavior measure and performance measure. A positive correlation indicates that the extent of environmental behavior and discharge ratio generally vary together in the same direction. A negative correlation indicates that the extent of environmental behavior and discharge ratio generally vary together in opposite directions. We also assess the magnitude of these correlations in order to explore the strength of the relationships between environmental behavior and performance. In addition, we assess the statistical significance of these correlations, as captured by the p-values associated with the correlations, in order to test whether the calculated magnitudes are distinguishable from zero, which implies no relationship between an individual environmental behavior measure and an individual environmental performance measure. We deem a correlation statistically significant when our analysis is able to reject the null hypothesis of a zero correlation at a significance level of 10 percent or less, as revealed by a p-value of 0.10 or less. We deem a correlation statistically insignificant when our analysis is not able to reject the null hypothesis of a zero correlation at a significance level of 10 percent or less, as revealed by a p-value greater than 0.10.[8]

Lastly, for our examination of the effects of environmental behavior on environmental performance, we must limit our data on environmental performance to a time frame that is identical to our data on environmental behavior. Recall that some behavioral forms are measured for each of three calendar years between 1999 and 2001. Other behavioral forms are measured for the twelve-month period preceding the completion of the survey. The remaining behavioral forms are measured for the three-year period preceding the completion of the survey. Since the time frame differs across three broad categories of behavioral forms, our exploration of the relationship between behavior and performance involves three different time frames.

6.7.3. Analytical Results and Interpretation

We explore a variety of environmental behavioral measures. First, we explore the quantity and quality of labor effort devoted to environmental management. We measure quantity based on two classifications: overall number of environmental employees and the number of employees devoted to wastewater management. In both cases, we explore the absolute number of employees and the ratio of environmentally or wastewater-related employees to the overall number of employees working at a particular facility. As shown in Table 6.3, facilities employing an above-median number of environmental employees significantly comply less fully with their TSS and BOD discharge limits than facilities employing a below-median number of environmental employees. In stark contrast, based on the correlation ($\rho = -0.092$, $p < 0.0001$), as facilities hire more environmental employees, the extent of compliance significantly improves (that is, discharge ratio falls). Perhaps the simple distinction between a smaller number and larger number of environmental employees masks the positive relationship between better behavior and better performance.

By relating the number of environmental employees to the overall number of facility employees, the analysis provides evidence to support further this positive relationship between labor devoted to environmental management and environmental performance. As shown in Table 6.3, facilities employing an above-median environmental employee to overall employee ratio comply more fully with their discharge limits than facilities employing a below-median ratio; however, the difference is statistically insignificant. As stronger evidence, based on the correlation ($\rho = -0.038$, $p = 0.038$), as facilities employ a greater ratio of environmental employees, the extent of compliance significantly improves (that is, discharge ratio falls).

In general, these results indicate that better behavior, as measured by a greater quantity of labor effort devoted to environmental management, seems to improve environmental performance.

An exploration of wastewater employees mostly supports this conclusion. As shown in Table 6.3, similar to environmental employees in general, facilities employing an above-median number of wastewater employees comply less fully with their discharge limits than facilities employing a below-median number of wastewater employees. This difference is significant based on only the median two-sample test ($p = 0.017$). In stark contrast, based on the correlation ($\rho = -0.072$, $p = 0.011$), as facilities hire more wastewater employees, the extent of compliance significantly improves (that is, discharge ratio falls). As with all types of environmental employees, perhaps the simple distinction between a smaller number and larger number of wastewater employees masks the positive relationship between better behavior and better performance.

By relating the number of wastewater employees to the overall number of facility employees, the analysis provides similar evidence to support this positive relationship. As shown in Table 6.3, facilities employing an above-median wastewater employee to overall employee ratio comply less fully with their discharge limits than facilities employing a below-median ratio, though the difference proves statistically insignificant. As stronger evidence, based on the correlation ($\rho = -0.066$, $p = 0.019$), as facilities employ a greater ratio of wastewater employees, the extent of compliance significantly improves (that is, discharge ratio falls).

We measure the quality of labor effort based on the employees' education and work experience. As shown in Table 6.3, facilities employing an above-median percentage of college-educated environmental employees more fully comply with their discharge limits than facilities employing a below-median percentage of college-educated environmental employees. This difference is statistically significant based on the median two-sample test ($p = 0.038$). Consistent with this significant difference, based on the correlation ($\rho = -0.103$, $p < 0.0001$), as facilities hire more college-educated

TABLE 6.3

Hypothesis testing: Effects of environmental behavior on environmental performance as measured by discharge ratio based jointly on TSS and BOD quantity and concentration discharges and limits

Behavioral Measure	Behavior-Based Subsamples: Lesser Behavior Versus Greater Behavior	N	Median	Sign of Difference in Medians[a]	Test Statistic (p-value)[b]
Environmental Employees: Absolute	Below median behavior	1281	0.183	+	2.611
	Above median behavior	1316	0.197		(0.005)
Environmental Employees: Relative to Overall Employees	Below median behavior	1318	0.197	–	–0.515
	Above median behavior	1231	0.193		(0.303)
Wastewater Employees: Absolute	Below median behavior	512	0.168	+	1.162
	Above median behavior	514	0.195		(0.123)
Wastewater Employees: Relative to Overall Employees	Below median behavior	523	0.176	+	1.014
	Above median behavior	490	0.193		(0.155)
Environmental Employees' Education	Below median behavior	496	0.191	–	0.437
	Above median behavior	504	0.168		(0.331)
Environmental Employees' Work Experience	Below median behavior	553	0.209	–	–3.846
	Above median behavior	447	0.159		(0.001)
Training: Quality	Lower: none, internal only, external only	589	0.212	–	–2.822
	Higher: internal and external	437	0.165		(0.002)
Training Days	Below median behavior	612	0.184	–	–2.240
	Above median behavior	349	0.153		(0.013)
Training Attendance	Below median behavior	242	0.213	–	–1.889
	At median behavior	758	0.174		(0.030)
TSS Treatment Presence	Absent	60	0.294	–	–1.811
	Present	953	0.174		(0.035)

Variable	Category	N	Value	Sign	Statistic (p-value)
TSS Treatment Extent	Lesser: none, coagulation+sedimentation, sedimentation in effluent pond	433	0.160	+	−3.476 (0.001)
	Greater: sedimentation in final clarifier, granular media filtration, membrane filtration	580	0.193		
BOD Treatment Presence	Absent	138	0.202	−	−0.808 (0.210)
	Present	888	0.177		
BOD Treatment Extent	Lesser: none, solids removal, stabilization pond, trickling filter	380	0.171	+	−1.228 (0.110)
	Greater: activated sludge, chemical oxidation, carbon adsorption	646	0.179		
Treatment Upgrade: Presence	No	910	0.131	+	11.500 (0.001)
	Yes	1852	0.221		
Treatment Upgrade: Designed to Reduce Wastewater	No	1253	0.151	+	10.779 (0.001)
	Yes	1435	0.241		
Monitoring Program	Lesser: none (except final discharge), treatment process, single-point sewer and treatment process	463	0.232	−	−4.496 (0.001)
	Greater: multiple-point sewer and treatment process	563	0.160		
Audit: Annual Count	Below median behavior	1310	0.192	−	−0.813 (0.208)
	Above median behavior	1134	0.191		
Audit: Team Composition	Lesser: none, visiting corporate/consultants, teams of facility employees	333	0.173	+	−0.046 (0.482)
	Greater: teams of visiting corporate/consultants and facility employees	641	0.187		

(continued)

TABLE 6.3 (*continued*)

Behavioral Measure	Behavior-Based Subsamples: Lesser Behavior Versus Greater Behavior	N	Median	Sign of Difference in Medians[a]	Test Statistic (p-value)[b]
Audit: Classification Protocol	Lesser: none, noncompliance findings, noncompliance versus other, priority categories of vulnerability	781	0.172	+	0.491 (0.312)
	Greater: ranked serially by vulnerability for noncompliance	221	0.224		
External Consultants: Use	No	665	0.224	–	-4.420 (0.001)
	Yes	2134	0.178		
External Consultants: Annual Expenditures—Absolute	Below median behavior	1437	0.208	–	-3.592 (0.001)
	Above median behavior	1183	0.168		
External Consultants: Annual Expenditures—Relative to Overall Employees	Below median behavior	1401	0.203	–	-3.575 (0.001)
	Above median behavior	1219	0.169		
Requested Assistance from Wastewater Regulator	No	2157	0.195	–	-1.698 (0.045)
	Yes	642	0.176		
Compliance Assistance Meets Facility's Environmental Employees' Needs	Lesser: never, some of time, most of time	1761	0.185	–	-4.371 (0.001)
	Greater: always	878	0.153		
Importance Management Places on Role of General Plant Employees in Identifying and Correcting Conditions Leading to Noncompliance	Lesser: somewhat important, important	181	0.117	+	5.366 (0.001)
	Greater: very important	845	0.199		
	At or above midpoint	514	0.181		

Management's Steps to Communicate Environmental Performance Goals and Targets to Facility Employees (count)	Below median	756	0.189	—	−2.299
	Above median	270	0.153		(0.011)
Management's Steps to Communicate Environmental Performance Goals and Targets to External Parties (count)	Below median	589	0.182	—	−0.429
	Above median	437	0.175		(0.334)
ISO 14001 Compliance	No	821	0.191	—	−0.914
	Yes	142	0.166		(0.180)

[a]The difference in median values = (median of greater behavior subsample) − (median of below lesser behavior subsample).

[b]Test statistic applies to the comparison between the greater behavior subsample and lesser behavior subsample; it reports the Wilcoxon two-sample test in the form of a normal approximation. The p-value stems from a one-tailed test.

environmental employees, the extent of compliance significantly improves (that is, discharge ratio falls).

In the case of work experience, facilities employing better-experienced environmental employees comply more fully with their TSS and BOD discharge limits, based on all of the analytical results. For example, the correlation indicates that, as facilities hire better-experienced environmental employees, the extent of compliance improves ($\rho = -0.038$) but significantly so only at a marginal level ($p = 0.117$). We also measure the quality of labor effort based on the provision of training, its quality, and its quantity. As shown in Table 6.3, facilities providing higher-quality training significantly comply more fully with their discharge limits than do facilities providing lower-quality training. Figure 6.4 (top) displays this effect also. As important, facilities providing an above-median number of training days significantly comply more fully with their discharge limits than do facilities providing a below-median number of training days. As further evidence, based on the correlation ($\rho = -0.002$, $p = 0.469$), as facilities provide more training days, the extent of compliance improves but not significantly. As the other dimension of training quantity, facilities inducing perfect attendance at their training sessions significantly comply more fully with their discharge limits than do facilities inducing less-than-perfect attendance. Consistent with this difference, based on the correlation ($\rho = -0.011$, $p = 0.370$), as facilities prompt better attendance, the extent of compliance improves but insignificantly so.

In general, these results indicate that better behavior, as measured by a greater quality of labor effort devoted to environmental management, seems to improve environmental performance.

Second, in contrast to labor effort, we next explore the provision of capital equipment. We explore the installation and use of treatment technologies and the extent of this treatment; we analyze TSS removal and BOD removal separately. We first contrast the presence of any pollutant-specific treatment technology with its absence and then distinguish between more effective treatment technologies and less effective technologies. As shown in Table 6.3, facilities employing any TSS treatment technology significantly comply more fully with their TSS and BOD discharge limits than facilities employing no TSS treatment technology. Figure 6.4 (middle) displays this effect more dramatically. Similarly, facilities employing any BOD treatment technology comply more fully with their TSS and BOD discharge limits than do facilities employing no BOD treatment technology, as shown in both Table 6.3 and Figure 6.4 (bottom); however, the difference is statistically insignificant. This difference proves highly significant when the analysis considers only BOD-related discharge ratios ($p < 0.0001$). Moreover, facilities employing more-effective TSS treat-

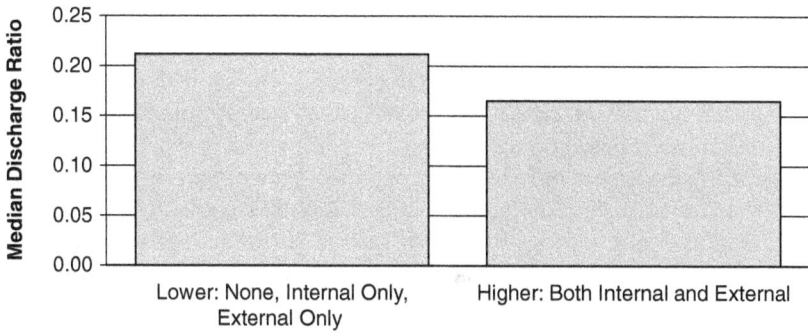

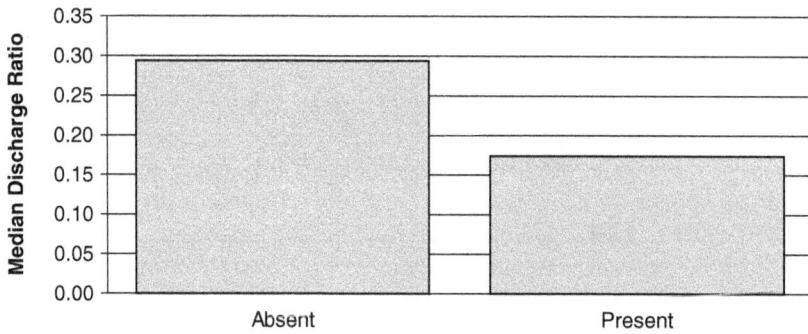

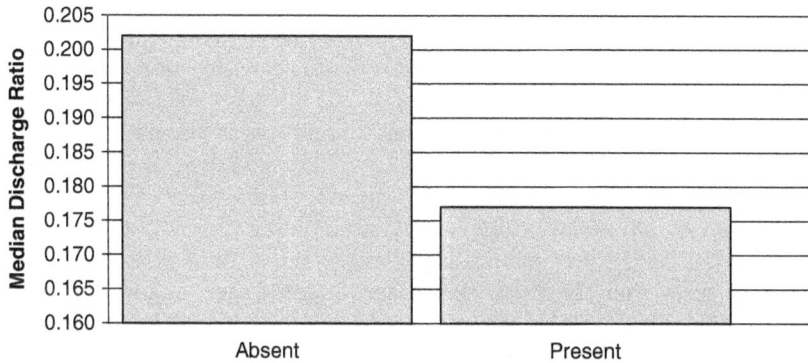

FIGURE 6.4

Effects of environmental behavior on discharge ratio.
Top: Quality of environmental training. Middle: TSS treatment presence.
Bottom: BOD treatment presence.

ment technologies significantly comply less fully with their discharge limits
than do facilities employing less-effective technologies.[9] A similar conclusion applies to BOD treatment technologies, but the difference is only
marginally significant based on the Wilcoxon two-sample test ($p=0.11$)
and insignificant based on the median two-sample test ($p=0.349$).[10] When
the analysis considers only BOD-related discharge ratios, the results indicate that facilities employing more-effective BOD treatment technologies
significantly comply more fully with their BOD limits than do facilities
employing less-effective BOD technologies ($p=0.078$ based on the median two-sample test). These results indicate that better behavior, as measured by the use and effectiveness of both TSS and BOD treatment
technologies, improves environmental performance. The evidence is
mixed in the case of TSS-related treatment technology and environmental performance.

We also assess upgrades to these treatment technologies based on the
presence of any type of upgrade and the presence of an upgrade designed
to reduce wastewater discharges. Regardless of the purpose of the upgrade,
facilities upgrading their treatment technologies significantly comply less
fully with their discharge limits than facilities not upgrading. Thus, better
behavior in the form of treatment upgrades appears to undermine environmental performance, a rather disturbing conclusion. Given these disturbing results, we are obligated to acknowledge that our analytical approach
is vulnerable to a situation of reverse causality. Rather than treatment upgrades causing better environmental performance, weak environmental
performance may be causing treatment upgrades. Using our data on the
timing of treatment upgrades, future analysis intends to examine whether
previous weak environmental performance prompts treatment upgrades.

Third, we assess facilities' efforts to understand their environmental
management concerns. We first assess the extent of a facility's monitoring
program. As shown in Table 6.3, facilities operating more extensive
monitoring programs significantly comply more fully with their discharge limits than facilities operating less extensive monitoring programs. Thus, better behavior in the form of more extensive monitoring
improves environmental performance.

As further information about facilities' efforts to understand their environmental management concerns, we assess facilities' effort to audit
their operations measured both in quantity and quality, which is based
on the audit team's composition and the extent of the audit classification
protocol. As shown in Table 6.3, facilities conducting an above-median
number of audits comply more fully with their discharge limits than do
facilities conducing a below-median number of audits, but the difference
is statistically insignificant. Similarly, based on the correlation ($\rho=-0.029$,

p=0.074), as facilities conduct more audits, the extent of compliance significantly improves. As for the quality of audit efforts, facilities utilizing higher-quality audit teams comply less fully with their discharge limits than do facilities utilizing lower-quality audit teams. However, the difference is statistically insignificant. Facilities utilizing more extensive audit classification protocols comply less fully with their discharge limits than do facilities utilizing less extensive audit classification protocols. This difference is statistically significant based on only the median two-sample test (p=0.004).

Overall, these results indicate that a greater quantity of audits seems to improve environmental performance, but higher-quality audits may undermine environmental performance.

Fourth, we assess facilities' efforts to seek insight from others regarding their environmental management efforts, for example, by hiring external consultants. As shown in Table 6.3, facilities hiring external consultants significantly comply more fully with their discharge limits than facilities not hiring external consultants. We enhance this conclusion by considering the amount of money spent on external consultants, in an absolute sense and relative to a facility's size. In both cases, facilities spending an above-median amount on external consultants significantly comply more fully with their discharge limits than facilities spending a below-median amount on external consultants. As further evidence, the correlations indicate that the extent of compliance improves as facilities spend more on external consultants ($\rho=-0.010$ and $\rho=-0.043$, respectively, for absolute expenditures and relative expenditures); this relationship proves statistically significant by relating absolute expenditures to the facility's size (p=0.014) but not otherwise (p=0.305).

In contrast to hired consultants, facilities may seek assistance and insight from their regulators. As shown in Table 6.3, facilities seeking regulatory assistance significantly comply more fully with their discharge limits than facilities not seeking assistance.

As part of this assessment, we explore how frequently compliance-assistance materials, including those from consultants and regulators, meet the day-to-day needs of facilities' environmental employees. As shown in Table 6.3, facilities whose compliance materials always met their day-to-day needs significantly comply more fully with their discharge limits than facilities whose compliance materials are less helpful.

In general, these results indicate that better behavior in the form of greater insight from others regarding a facility's environmental management efforts leads to better performance.

Fifth, we assess indicators of facilities' degree of concern over environmental management and to what extent these concerns translate into goals that are communicated to the facility's employees. As shown in Table 6.3, facilities possessing a greater concern significantly comply less fully with their discharge limits than facilities possessing a lesser concern. Thus, better behavior in the form of greater environmental management concern appears to undermine environmental performance, indicating that good intentions impede good outcomes.

We also assess the degree of effort expended by management to communicate environmental performance goals and targets to facility employees, measured by the number of different categories of steps taken by management. As shown in Table 6.3, facilities taking an above-median number of steps significantly comply more fully with their discharge limits than facilities taking a below-median number of steps. As further evidence, based on the correlation, as facilities take more steps, the extent of compliance significantly improves ($\rho = -0.137$, $p < 0.0001$). Thus, better behavior in the form of greater internal communication of environmental goals improves environmental performance.

Sixth, in contrast to the communication to facility employees, we also assess the communication of environmental performance goals to targets to external parties, measured by the number of different categories of steps taken by management. As shown in Table 6.3, facilities taking an above-median number of steps comply more fully with their discharge limits than facilities taking a below-median number of steps, but the difference is statistically insignificant. As stronger evidence, based on the correlation, as facilities take more steps, the extent of compliance significantly improves ($\rho = -0.041$, $p = 0.096$). Again, greater communication improves environmental performance.

Seventh, we assess a broad indicator of a well-organized environmental management system: compliance with ISO 14001 standards. As shown in Table 6.3, ISO-compliant facilities comply more fully with their discharge limits than facilities lacking ISO compliance; this difference is statistically significant based only on the median two-sample test ($p = 0.104$) and even then only marginally so. Thus, better behavior in the form of ISO 14001 compliance leads to better performance.

Considering all measures of environmental behavior, while the evidence is clearly mixed, better behavior in general appears to lead to better environmental performance.

6.8. COMPARING ENVIRONMENTAL PERFORMANCE:
SURVEY RESPONDENTS VERSUS
SURVEY NONRESPONDENTS

In Chapter 3 we provide evidence to support the claim that the survey does not appear to suffer from nonresponse bias. In this section, we provide further evidence to support that claim. Since data on wastewater discharges is publicly available for all facilities permitted within the National Pollutant Discharge Elimination System, we are able to compare the environmental performance of facilities that responded to our survey ("survey respondents") and the environmental performance of facilities that we contacted as part of our survey but that failed to respond ("survey nonrespondents"). If the two groups of facilities generate comparable levels of environmental performance, then the results of our hypothesis testing—the effects of discharge limits on environmental performance and the effects of environmental behavior on environmental performance—are more likely to generalize from survey respondents to survey nonrespondents.

In order to assess whether the levels of environmental performance are comparable, we employ two-sample tests. First, we identify each group of facilities as a separate subsample: (1) survey respondents subsample, and (2) survey nonrespondents subsample. Given these two subsamples, we describe the distribution in each subsample and then compare the distributions. Finally, we employ three two-sample tests —the two-sample means t-test, the Wilcoxon two-sample test, and the median two-sample test[11]—in order to assess whether any identified difference in the subsample-specific distributions is statistically significant.[12]

This chapter explores various measures of environmental performance. Even though part of our hypothesis testing involves absolute discharges, we do not compare absolute discharges between survey respondents and survey nonrespondents because discharge limits may differ between the two groups of facilities. While we could test this concern, we instead choose to assess the measure of environmental performance that combines these two elements: the ratio of absolute discharges to discharge limits (that is, discharge ratio). Even though part of our hypothesis testing distinguishes between quantity-based compliance and concentration-based compliance, we do not compare measurement-specific compliance between survey respondents and survey nonrespondents because the imposition of quantity-based limits and concentration-based limits may differ between the two groups of facilities. While we could test this concern, we instead choose to assess the measure of discharge ratio that combines these two elements: quantity-based and concentration-based discharge ratios jointly considered. We assess separately TSS-related discharge ratios and BOD-related discharge ratios.

Consider first TSS-related quantity-based and concentration-based discharge ratios. The average respondent facility complies strongly with its TSS discharge limit given a mean discharge ratio of 0.267. The average nonrespondent facility complies identically given the same mean discharge ratio of 0.267. The two subsample-specific median values are nearly identical: the respondent facility median equals 0.192 and the nonrespondent facility median equals 0.194. Based on this evidence alone, environmental performance appears similar between the two groups of facilities. The two-sample means t-test statistics support this conclusion by failing to reject the null hypothesis of equal means (p=0.969).[13] As further evidence, the median two-sample test statistic fails to reject the null hypothesis of identical distributions (p=0.714). In contrast, the Wilcoxon two-sample test statistic rejects the null hypothesis of identical distributions (p=0.021), revealing that respondent facilities comply more fully with their TSS limits than nonrespondent facilities (that is, the distribution of the respondent subsample is located at lower TSS-related discharge ratios than the distribution of the nonrespondent subsample). In general, these results seem to indicate that TSS-related environmental performance does not differ between survey respondents and survey nonrespondents, though the evidence is mixed.

Consider next BOD-related quantity-based and concentration-based discharge ratios. The average respondent facility complies strongly with its BOD discharge limit given a mean discharge ratio of 0.261. The average nonrespondent facility also complies strongly given a mean discharge ratio of 0.256. These mean values are nearly identical. The two subsample-specific median values are quite similar: the respondent facility median equals 0.162 and the nonrespondent facility median equals 0.175. (Using the smaller median as the base, in order to maximize the difference between the two values, the median nonrespondent facility is 8.2 percent less compliant than the median respondent facility.) Based on this evidence alone, environmental performance appears similar between the two groups of facilities. As with TSS-related performance, the two-sample means t-test statistics support this conclusion by failing to reject the null hypothesis of equal BOD-related means (p=0.616).[14] However, the other two test statistics reject the null hypothesis of identical distributions, revealing that respondent facilities comply more fully with their BOD limits than nonrespondent facilities; that is, the distribution of the respondent subsample is located at lower BOD-related discharge ratios than the distribution of the nonrespondent subsample. (The p-values for the Wilcoxon two-sample test and the median two-sample test are 0.001 and 0.004, respectively.) Even though two of the three tests indicate that BOD-related environmental performance significantly differs between survey respondents and survey nonrespondents in a statistical sense, we argue that the magnitude of the difference seems insubstantial.

An assessment of discharge ratios based on both TSS and BOD discharges generates the same general conclusion: environmental performance does not differ meaningfully between survey respondents and survey nonrespondents.

Overall, these results seem to indicate that the hypothesis test results may generalize from survey respondents to survey nonrespondents.

6.9. SUMMARY

This chapter inquires whether imposed discharge limits affect environmental performance, as measured by both absolute discharges and absolute discharges relative to permitted discharges (that is, discharge ratio). We find that the median facility in our study was 80 percent overcompliant with TSS and BOD limits in general, and that the average facility was noncompliant in only 3.08 percent of the months in the sample period. Our first hypothesis is that absolute discharges fall as limits tighten (or absolute limits rise as limits loosen). If this relationship is true, then discharge limits appear effective at prompting better environmental performance. Our second hypothesis is that discharge ratios rise as limits tighten (or discharge ratios fall as limits loosen). If this relationship is true, then discharge limits appear not only effective but also binding (that is, facilities are constrained by the limits). Our results generally support the conclusions that both TSS and BOD discharge limits are effective, but that only BOD discharge limits are definitively binding because TSS-concentration limits appear to induce facilities to discover cheaper means of controlling TSS discharges. As our third hypothesis, we test whether better environmental behavior leads to better environmental performance. We find that in general, various forms of better behavior—including a greater quantity and quality of labor effort devoted to environmental management, more extensive monitoring, a greater quantity of audits, greater insight from others regarding a facility's environmental management, greater internal communication of environmental goals, and ISO 14001 compliance—seem to improve environmental behavior. However, treatment upgrades and higher-quality audits seem to undermine environmental performance. Overall, despite mixed evidence, better behavior appears generally to lead to better environmental performance.

Chapter Six Appendix

This appendix explores whether and to what extent environmental performance varies across a variety of dimensions. Facilities operating in different states or EPA regions may differ systematically in the extent of their success at controlling their discharges. Similarly, facilities operating in different sectors may differ systematically in their environmental success. As important, larger facilities may systematically differ from smaller facilities in their abilities, their motivations, or both, to control successfully wastewater discharges. We also explore differences over time; environmental performance may show a positive or negative trend. As a complement to an assessment of overall trends, we also explore trends within individual facilities by analyzing variation over time for each specific facility and summarizing these individual trends across all of the sampled facilities. We also explore the volatility of discharges for each individual facility. For this assessment, we explore only the most comprehensive measure of environmental performance for each pollutant: quantity- and concentration-based discharge ratio.

6A.1.1. Subsamples Based on Location: EPA Regions

As the first set of subsamples, we explore variation across facility location. In particular, we assess differences in environmental performance across EPA regions. Again, we identify the two most prominent regions in our sample—Region 4 and Region 6, while grouping the remaining regions into a single "other region" category. Table 6A.1 displays the regional variation. For example, based on either TSS or BOD discharge ratios, the median values reveal that facilities operating in Region 4 are

TABLE 6A.1

Environmental performance: Subsamples based on facility location—EPA region

Discharge Ratio: Concentration and Quantity (jointly considered)	Region[a]	N	Mean	Median
TSS Discharges	0	1,173	0.242	0.174
	4	1,253	0.286	0.203
	6	1,270	0.271	0.195
BOD Discharges	0	1,152	0.205	0.126
	4	965	0.263	0.205
	6	975	0.323	0.158

SOURCE: Environmental Protection Agency Permit Compliance System database.
[a]Region 0 refers to "other regions."

the least compliant, while facilities operating in "other regions" are the most compliant.

6A.1.2. Subsamples Based on Chemical Manufacturing Subsectors

We next assess differences in environmental performance across the three broad subsectors: organic chemicals, inorganic chemicals, and other chemicals. As part of this assessment, we test whether any noted difference between any two of the three subsectors is statistically significant. For example, we test whether a facility producing an inorganic chemical achieves a degree of success at controlling wastewater discharges that differs significantly from that of a facility producing an organic chemical. For each pairwise comparison, we identify one sample for each subsector and then employ two-sample tests: the Wilcoxon two-sample test and the median two-sample test. We report only the former test results since the latter test results lead to identical conclusions.[1]

Table 6A.2 displays the sectoral variation. Based on TSS discharge ratios, facilities producing organic chemicals are the least compliant, while facilities producing other chemicals are the most compliant, with facilities producing inorganic chemicals in between. Each of the pairwise comparisons proves statistically significant. In contrast, based on BOD discharge ratios, the median values reveal that facilities producing inorganic chemicals are the least compliant, while facilities producing other chemicals are the most compliant, with facilities producing organic chemicals in between. The two-sample tests reveal that other chemical facilities are significantly more compliant than either organic chemical

TABLE 6A.2
Environmental performance: Subsamples based on facility sector

Discharge Ratio: Concentration and Quantity (jointly considered)	Sector[a]	N	Mean	Median	I Versus G[a]	I Versus O[a]	G Versus O[a]
TSS	I	854	0.221	0.165	−7.7113	−5.0727	−11.5941
Discharges	G	2263	0.308	0.231	(0.0001)	(0.0001)	(0.0001)
	O	579	0.171	0.123			
BOD	I	431	0.223	0.179	0.2463	5.3689	−6.7866
Discharges	G	2115	0.288	0.174	(0.8054)	(0.0001)	(0.0001)
	O	546	0.186	0.089			

SOURCE: Environmental Protection Agency Permit Compliance System database.
[a]Sector abbreviations: I, inorganic chemical; G, organic chemical; O, other chemical.
[b]Test statistic applies to pairwise comparison of the three sectors; it reports the Wilcoxon two-sample test in the form of a normal approximation. The p-value is two-sided.

facilities or inorganic chemical facilities, but organic chemical facilities and inorganic chemical facilities display statistically similar environmental performance.

6A.1.3. Subsamples Based on Facility Size

In this section, we assess differences in environmental performance between smaller facilities and larger facilities. We split the sample into two subsamples based on the sample median facility size, as measured by the overall number of employees working at an individual facility. Based on this division, we then assess whether environmental performance differs between smaller facilities (that is, below median size) and larger facilities (that is, above median size) using two-sample tests: Wilcoxon two-sample test and median two-sample test. We tabulate only the former test results.[2]

Table 6A.3 displays the means, medians, and test statistics. Based on both TSS- and BOD-related environmental performance, larger facilities generate better performance than do smaller facilities, as evidenced by the comparisons of subsample-specific means and medians. These comparisons prove statistically significant based on the median two-sample test (p = 0.0003 and p = 0.0001 for TSS and BOD, respectively) and based on the Wilcoxon two-sample test in the case of BOD-related environmental performance.[3]

TABLE 6A.3

Environmental performance: Subsamples based on facility size

Discharge Ratio: Concentration and Quantity (jointly considered)	Size	N	Mean	Median	Test Statistic (p-value)[a]
TSS Discharges	Below median	1,797	0.292	0.208	1.1873
	Above median	1,899	0.243	0.179	(0.2351)
BOD Discharges	Below median	1,525	0.300	0.182	2.8889
	Above median	1,567	0.223	0.143	(0.0039)

SOURCE: Environmental Protection Agency Permit Compliance System database.

[a]Test statistic applies to the comparison between the below-median-size subsample and above-median-size subsample; it reports the Wilcoxon two-sample test in the form of a normal approximation. The p-value is two-sided.

TABLE 6A.4

Environmental performance: Subsamples based on year

Discharge Ratio: Concentration and Quantity (jointly considered)	Year	N	Mean	Median
TSS Discharges	1999	816	0.283	0.198
	2000	869	0.275	0.206
	2001	900	0.260	0.179
	2002	893	0.250	0.180
	2003	218	0.268	0.215
BOD Discharges	1999	715	0.241	0.158
	2000	724	0.253	0.164
	2001	738	0.250	0.143
	2002	735	0.293	0.174
	2003	180	0.283	0.174

SOURCE: Environmental Protection Agency Permit Compliance System database.

6A.1.4. Subsamples Based on Time: Years (1999, 2000, 2001, 2002, 2003)

In this section, we assess differences in environmental performance over time, as shown in Table 6A.4. Based on either year-specific means or year-specific medians, TSS-related environmental performance varies very little over time, with no obvious trend. The same conclusion applies to BOD-related environmental performance. Even though one might hope that technological advances and the growing prevalence of corporate environmental sustainability practices improve environmental performance over time, the reported data do not support this hope.

6A.2. BEHAVIORAL VARIATION OVER TIME FOR
A GIVEN FACILITY

Finally, we explore variation in environmental performance over time for a given facility. First, we explore trends within individual facilities by calculating the correlation between each specific facility's environmental performance and a time trend (that is, an index that equals one in the first month of the sample period—January 1999—and equals 51 in the last month of the sample period—March 2003). Second, we explore the volatility of discharges within an individual facility by calculating each facility's mean quantity- and concentration-based discharge ratio over the sample period and each specific facility's standard deviation about this mean. (The standard deviation equals the square root of the calculated variance about the mean, which equals the sum of squared deviations from the mean across all the relevant observations; that is, for each observation, calculate the deviation from the sample mean, square this deviation, and then sum across all the relevant observations.) We then divide the standard deviation by the mean to generate a deviation-mean ratio.

Based on TSS-related environmental performance, the average facility generates no trend in its performance given a correlation with the time trend of −0.009. The median facility generates a comparably nonexistent trend given a correlation of −0.024. Nevertheless, upward of 25 percent of the sample generates a meaningful improvement in performance; that is, the individual facility-specific correlation lies at or below −0.247 for 25 percent of the facilities. In contrast, at least 25 percent of the sample unfortunately generates a meaningful degradation in performance; that is, the individual facility-specific correlation lies at or above 0.248 for 25 percent of the facilities. Similarly, based on BOD-related discharge ratios, the average facility generates no substantive trend in its performance given a correlation with the time trend of −0.026. The median facility generates a comparably weak trend given a correlation of −0.016. Nevertheless, upward of 25 percent of the sample generates a meaningful improvement in performance; that is, the individual facility-specific correlation lies at or below −0.390 for 25 percent of the facilities. In contrast, at least 25 percent of the sample unfortunately generates a meaningful degradation in performance; that is, the individual facility-specific correlation lies at or above 0.233 for 25 percent of the facilities.

Second, we explore the volatility of discharge ratios, as measured by the deviation-mean ratio. Based on TSS-related environmental performance, compliance is rather volatile. The median facility generates a standard deviation that equals 59 percent of its mean level. Even a facility near the bottom of the distribution (tenth percentile) generates a standard deviation that equals 36 percent of the facility-specific mean level. As important, a facility near the top of the distribution (ninetieth percentile) generates much volatility, given a standard deviation that equals 146 percent of the mean level. A similar conclusion stems from an assessment of BOD-related environmental performance. The median facility generates a standard deviation that equals 62 percent of its mean level. Meaningful volatility is also found at the lower and upper reaches of the distribution: a facility near the bottom of the distribution (tenth percentile) generates a standard deviation that equals 36 percent of the facility-specific mean level, while a facility near the top of the distribution (ninetieth percentile) generates a standard deviation that equals 126 percent of the facility-specific mean level.

Regulatory Efforts to Induce
Compliance with Discharge Limits

This chapter explores regulators' attempts to induce compliance with wastewater discharge limits. As the key purpose of this exploration, the chapter attempts to answer the following research question: what are regulators doing to induce environmental behavior that effectively controls discharges so that the discharges comply with the imposed limits? Specifically, we seek to answer this question: what government interventions, namely, inspections and enforcement actions, are regulators taking in order to induce compliance?

7.1. THE ENVIRONMENTAL PROTECTION AGENCY'S
ENFORCEMENT AND MONITORING AUTHORITY

Regulatory restrictions, such as discharge limits, applicable to individual facilities are meaningful only if dischargers comply with them. Dischargers have greater incentives to comply (and more to fear if they do not) if regulators have the capacity to impose unwanted consequences on those who do not comply. As Jean-Jacques Rousseau observed, if people and their property "were not answerable for personal actions, nothing would be easier than to evade duties and laugh at the laws" (Rousseau, 1755, p. 138).

The Clean Water Act provides the Environmental Protection Agency (EPA) with a substantial arsenal of weapons to investigate suspected violations and initiate enforcement actions on the basis of the information revealed during investigations. This arsenal includes a variety of weapons to bring noncomplying facilities into compliance and penalize past noncompliance, particularly in the pocketbook. State environmental agencies administering the National Pollutant Discharge Elimination System (NP-DES) permit program must have similar options under their own laws as a prerequisite to being allowed to achieve primacy in implementation of that program.[1] This chapter summarizes the civil enforcement mechanisms used to enforce the Clean Water Act. It then discusses the inspections and enforcement actions brought by EPA and the inspections brought by the states against facilities in the chemical industry during the period covered by our study.[2]

7.2. INSPECTIONS AND ENFORCEMENT ACTIONS

Successful enforcement of any laws, including environmental laws, depends on the accumulation of information that provides proof that a violation has occurred. The Clean Water Act authorizes federal and state regulators seeking to determine the compliance status of regulated facilities to gather information through facility inspections. The statute also allows regulators to require facilities to file monthly discharge-monitoring reports and maintain additional records. The records that discharging facilities must maintain may include information concerning an analysis of the chemical constituents of any discharge; appropriate bioassays necessary to determine the concentrations of regulated pollutants present in wastewater flow and an analysis of initial dilution; available process modifications that reduce discharges; analysis of the location where pollutants are sought to be discharged; and evaluation of available alternatives to the discharge of the pollutants, including an evaluation of the possibility of land-based disposal.[3] If regulators detect violations, they may initiate enforcement actions against the responsible facilities. The following sections provide greater detail on the role of both inspections and enforcement actions in the identification and correction of Clean Water Act violations.

7.3. STATE AND FEDERAL INSPECTIONS

As the first element of ensuring compliance with Clean Water Act provisions, inspections provide a tool for assessing a facility's compliance. EPA has the right under the Clean Water Act to enter any premises in which an effluent source is located. It also has the right to enter any premises in which any records that facilities are required to maintain are kept for the purpose of determining whether the source is in violation of NPDES permit provisions, including applicable discharge limits. The agency may at reasonable times have access to and copy any records, inspect any monitoring equipment, or sample any effluents that the facility is required to sample.[4] If EPA finds that state laws and procedures relating to inspection, monitoring, or entry apply to the same extent as those required of EPA under the Clean Water Act, the state is authorized to apply and enforce those laws and procedures with respect to facilities located in the state.[5] State agencies administering the NPDES program have similar authority.

EPA and the states conduct different kinds of inspections, depending on the purpose of the inspection. All major facilities must be inspected at least annually. The type of inspection is left to the discretion of the appropriate EPA regional or state office, however (EPA, 1990). In the case of minor facilities, both the number and identity of facilities that are inspected annually also are left to the discretion of the appropriate EPA regional or state office (EPA, 1990). Inspections differ widely with regard to the type of information collected and the vigor of the inspection oversight. Some inspections, such as reconnaissance inspections, involve only a brief visual inspection of the permitted facility, effluents, and receiving waters in order to obtain a preliminary overview of the permit holder's compliance program and produce a quick summary of potential compliance problems. Performance audit inspections and compliance evaluation inspections, which we describe in Section 7.6.4 later in this chapter, are more resource intensive (Helland, 1998b). EPA inspections usually occur without prenotification of the exact date, and notification is not recommended when EPA suspects illegal discharges or improper records (EPA, 1994).

7.4. AVAILABLE ENFORCEMENT ACTIONS

The Clean Water Act authorizes EPA to undertake a variety of enforcement actions, ranging from notification of an alleged violation to the pursuit of formal sanctions in either administrative or judicial forums. The Clean Water Act subjects violators to civil and criminal penalties, but this book deals only with civil, not criminal enforcement.[6] State environmental agencies must have similar authority in order to receive EPA's permission to administer the NPDES permit program.

EPA distinguishes between formal and informal enforcement actions. The difference between the two is that informal responses are without legal force and are designed to bring violators into compliance. Formal responses have legal force, which is why they are accompanied by procedural safeguards (such as prior notice and, in some cases, a trial-type procedure involving examination and cross-examination of witnesses) to afford regulated facilities an opportunity to defend themselves against charges of regulatory violations (Rechtschaffen, 2004).

As EPA uses the term, a "formal" action is one that "requires actions to achieve compliance, specifies a timetable (schedule), contains consequences for noncompliance that are independently enforceable without having to prove the original violation, and subjects the person to adverse legal consequences for noncompliance" (EPA, 1986). These formal actions include administrative compliance orders, cease-and-desist orders, settlement agreements (if they include measures required to return a facility to compliance or a schedule for taking actions needed to achieve compliance), and judicial actions that culminate in fines. Formal actions also include either administrative or judicial actions that culminate in the issuance of injunctions or the imposition of supplemental environmental project requirements.

Informal actions may take the form of telephone calls or other oral discussions with facility operators about compliance status, site visits, warning letters, or the issuance of a notice of violation. In a notice of violation, EPA or a state agency communicates to the permitted facility that it has detected a violation, which the permit holder should correct in order to avoid the initiation of formal actions (EPA, 1990).

EPA and the states often use informal actions to address what they regard as lower-priority violations or to deal with first-time violators. Enforcement officials may use the information they acquire in pursuing informal action if they later decide to engage in more aggressive enforcement (Silverman, 1990). Hunter and Waterman (1992) find that about 70 percent of the enforcement actions taken by EPA's water-pollution officials are informal in nature. Among enforcement actions taken by the states, informal actions also seem to predominate (Zinn, 2002).

7.4.1. Fines

Monetary fines, whether imposed through an administrative order or by a court, have the potential to induce regulated facilities to change their behavior in ways that improve environmental performance. Fines have the potential to increase the direct costs attributable to noncompliance, thereby creating incentives to avoid noncompliance that may lead to the imposition of fines. EPA has stated its confidence in the utility of monetary sanctions as an effective deterrent to noncompliance.

The Clean Water Act imposes limits on the amounts of monetary penalties recoverable from violators in both civil and administrative proceedings. Any facility that violates its discharge limits is subject to a judicially imposed civil penalty not to exceed $25,000 for each day of violation.[7] In determining the amount of a civil penalty to assess against a facility after a finding of violation, courts must consider factors specified in the statute, including the seriousness of the violation, the economic benefit (if any) resulting from the violation, a prior history of violations, any good-faith efforts to comply with applicable discharge limits, the economic impact of the penalty on the violator, and an open-ended category of "such other matters as justice may require."[8]

When EPA seeks to impose monetary penalties in administrative proceedings, the Clean Water Act distinguishes between two classes of monetary penalties. Administrative penalties for Class I violations, which govern minor violations (a term that is not defined in the statute), may not exceed $10,000 per violation, and the maximum total amount of any Class I penalty may not exceed $25,000. Administrative penalties for Class II violations, which are appropriate for more egregious violations (again, a term not defined in the Clean Water Act), may not exceed $10,000 per day for each day during which the violation continues, while the maximum total amount of any Class II penalty may not exceed $125,000.[9] According to EPA, the statutory maximum penalty for violations of a discharge limit for a period longer than one day includes a separate penalty for each day in the time period (assuming there was a discharge on each day). Because more is at stake for the discharger that is the target of enforcement in a Class II proceeding, the procedural rights afforded to alleged violators are more extensive than they are in Class I proceedings.

The factors relevant to determining the appropriate civil penalty assessment are similar to the ones that govern penalty calculations in judicial actions. The Clean Water Act provides that in determining the amount of any penalty to impose, EPA must take into account the nature, circumstances, extent, and gravity of the violation, and the economic benefit the violator gained through the violation.

EPA has provided insight into the meaning and relevance of many of these factors in documents it has issued to guide its own officials when they consider whether to settle pending enforcement actions. In deciding whether to settle pending enforcement cases, either in court or in administrative enforcement cases, EPA officials are supposed to determine the appropriate penalty to impose based on the following formula:

Penalty = Economic Benefit + Gravity +/– Gravity Adjustment Factors
    – Litigation Considerations – Ability to Pay
    – Supplemental Environmental Projects

The rationale for attempting to extract from the violator the economic benefit it received from noncompliance is to place the violator in the same financial position as it would have been if it had complied with applicable discharge limits on time. Facilities whose discharges exceed these limits are likely to have obtained an economic benefit as a result of delayed or completely avoided pollution control expenditures during the period of noncompliance. These delayed or avoided expenditures might include, for example, the costs of monitoring and reporting (including the costs of sampling and conducting proper laboratory analysis); delayed or avoided capital equipment improvements or repairs; and delayed or avoided operation and maintenance expenses.

The standard method that EPA uses to calculate the economic benefit from delayed and avoided pollution control expenditures is to apply its BEN model. EPA relies on BEN to generate "a reliable, objective dollar figure" that represents the economic benefit a violator gained through its violations (Markell, 2007, p. 562). EPA has not specified a minimum amount that triggers the use of the BEN model. In estimating a violator's economic benefit using the BEN model, EPA generally does not go back more than five years before the date a complaint was filed.[10]

In deciding whether to settle an administrative proceeding or civil action against an alleged violator, EPA also determines the amount of civil fines for which it is willing to settle on the basis of the gravity of the violation. EPA's reasoning is that removing the economic benefit of noncompliance only places the violator in the same position it would have been in if compliance had been achieved on time. According to EPA, "Both deterrence and fundamental fairness require that the penalty include an additional amount to ensure that the violator is economically worse off than if [it] had obeyed the law" (EPA, 1995, p. 6). The gravity component of the penalty is calculated for each month in which there was a violation. The total gravity component for the penalty calculation equals the sum of each monthly gravity component. The monthly gravity formula is the following:

$$\text{Monthly gravity component} = (1 + A + B + C + D) \times \$1,000.$$

Factor A is based on the degree to which discharges exceeded applicable discharge limits. Factor B concerns the degree to which the discharge caused actual or potential harm to human health or the environment. Factor C is based on the total number of effluent limit violations each month. Factor D involves the significance of violations of requirements other than discharge limits (which are described immediately below). This factor has a value ranging from 0 to 70 and is based on the severity and number of the different types of noneffluent limitation requirements violated each month. These other kinds of violations include monitoring and reporting, sludge handling, and noncompliance, with interim deadlines for taking specific actions (such as constructing new treatment facilities). EPA may adjust the gravity factor based on three factors: flow reduction factor (to reduce gravity); history of recalcitrance (to increase gravity); and the quick-settlement reduction factor (to reduce gravity).

7.4.2. Supplemental Environmental Projects

Another formal enforcement action is the supplemental environmental project (SEP). Although the Clean Water Act does not mention SEPs, they have been used in both administrative and judicial enforcement proceedings. SEPs are environmentally beneficial projects that a court or a regulatory agency allows a violator to implement in order to avoid a fine or to reduce the size of a fine that would otherwise be appropriate (Rechtschaffen and Markell, 2003). These might take the form of agreements to repair damage caused by the facility's previous discharges, create new or preserve existing wetlands to improve water quality, construct facilities to reduce storm-water runoff, or purchase emergency response equipment for the local government. EPA has authorized SEPs as part of the process of settling administrative complaints, which are formal charges filed by enforcement authorities in an administrative rather than a judicial proceeding (EPA, 2001). A facility that agrees to perform a SEP commits to taking action beyond the steps that are necessary to correct the identified violations. EPA provides oversight to ensure that the company does what it has promised to do (EPA, 2001). According to some observers, however, SEPs may actually lower the cost to regulated entities of regulatory violations, which may result in "underdeterrence of regulatory violations where there previously was none and worsen underdeterrence where there previously was some" (Dana, 1998, p. 1184).

7.4.3. Injunctive Relief

Yet another formal enforcement action is the issuance of injunctive relief. Injunctive relief refers to an order requiring that a regulated facility take specific action as a means of coming into compliance with a discharge limit with which a facility is not complying at the time of issuance. Typically, an injunction "should simply order the defendant to do what law-abiding citizens in the same situation would do voluntarily. That order might necessitate the immediate cessation of the offending conduct, depending on congressional intent" (Farber, 1984, p. 415). Injunctive relief can take the form of an administrative order issued by EPA or a state environmental agency or a decree issued by a civil court. Administrative orders mandating specific steps for the violator to come into compliance should include schedules by which each specific mandated action must be taken, known as "compliance schedules" (EPA, 1986).

EPA maintains the Permit Compliance System (PCS) database partially in order to record inspections and enforcement actions. First, the PCS database records inspections taken by EPA regional offices and state environmental agencies with Clean Water Act primacy. The PCS database also records all informal enforcement actions and most formal enforcement actions taken by EPA regional offices or imposed by civil courts with the assistance of the Department of Justice. The PCS database includes all federal formal actions that are neither fines, nor injunctive relief sanctions, nor SEPs, such as compliance orders. While the PCS database records federal fines imposed administratively or judicially, this database excludes federally imposed injunctive relief sanctions and SEPs. As a supplement to the PCS database, the EPA Docket database provides information on federal SEPs and injunctive relief sanctions through 2001. However, the successor to the Docket database—the EPA Integrated Compliance Information System database—does not provide this information in any usable manner. For this reason, our assessment of federal SEPs and injunctive relief sanctions ends in 2001. (Due to our lack of foresight regarding the replacement of the Docket database with the Integrated Compliance Information System database, our onetime extract from the Docket database did not include complete data for the year 2001; consequently, we must end our assessment of federal SEPs and injunctions in June of 2001.) More important, none of these databases systematically record enforcement actions taken by state environmental agencies with Clean Water Act primacy. While the PCS database provides some data on state enforcement actions, the authors confirmed that the available state enforcement data are quite incomplete by comparing the PCS data with extracts taken directly from the four state environmental agencies operating in Texas, Louisiana, New Jersey, and West Virginia. These four states represented the states with the largest number of chemical manufacturing facilities permitted as major dischargers within the NPDES system during our sample period.

7.6. USE OF INSPECTIONS

Next, we describe the use of government interventions within our chosen sample period, starting with inspections. For this description, we explore inspections taken against all Clean Water Act–regulated major facilities operating in the chemical manufacturing sector during our sample period, rather than exploring survey respondents exclusively, for two reasons. First, this broader depiction facilitates a richer understanding of the use of inspections. Second, in order to assess the effect of general deterrence stemming from inspections on the surveyed facilities' environmental behavior and performance, as further explained in Chapter 8, we needed to gather data on this broader set of facilities. We explore inspections only against major facilities because our survey respondents are all major facilities. As noted in Section 7.3, major facilities should be inspected annually, unlike minor facilities.

In addition to our depiction of all Clean Water Act–regulated major facilities, we also describe the inspections taken against only the surveyed facilities. In this way, our depiction of inspections is comparable to our previous depictions of the surveyed facilities. Moreover, this depiction is useful for understanding better our measures of specific deterrence, which are further explained in Chapter 8. When depicting use of inspections in the full set of Clean Water Act–regulated chemical manufacturing facilities, we divide the sample into subsamples according to location (that is, EPA region) and time (for example, calendar year). When depicting use of inspections in the set of surveyed facilities, we divide the sample into subsamples according to the chemical manufacturing subsector and facility size. We do not possess data on these two latter facility characteristics for all Clean Water Act–regulated facilities. Lastly, we expand our sample window to include the year 1998 because our measures of specific deterrence utilize inspections taken in this year, as explained in Chapter 8.

7.6.1. Use of Inspections for Entire Sample

We first describe inspections for the full set of Clean Water Act–regulated major chemical manufacturing facilities and then for only the surveyed facilities. During the expanded sample period between January 1998 and March 2003, 536 chemical manufacturing facilities were permitted within the NPDES system at some point. Over this time period, EPA regional offices performed 179 inspections at these 536 facilities. State environmental agencies performed 3,249 inspections. At the 97 survey respondents, EPA performed 34 inspections and state agencies performed 750 inspections.

7.6.2. *Use of Inspections Across Various Subsamples*

Next we describe how the use of inspections varies across space (for example, EPA regions), time, subsectors within the chemical manufacturing sector, and different facilities based on size. First, inspections vary across EPA regions. Based on the broad set of 536 Clean Water Act–regulated major facilities, at the 133 major facilities located within Region 4, EPA performed 28 inspections, while state agencies performed 896 inspections. At the 171 major facilities located within Region 6, EPA performed 57 inspections, while state agencies performed 589 inspections. At the 232 major facilities located within the other EPA regions, EPA performed 94 inspections, while state agencies performed 1,764 inspections. (The appendix to this chapter assesses the use of inspections in individual regions for certain years.)

The use of inspections also varies over time between 1998 and 2002. Based on the broad set of 536 Clean Water Act–regulated major facilities, the number of EPA inspections drops from a high of 43 in 1999 (with a comparable number of 42 in 1998) to a low of 21 in 2002. In between, EPA inspections total 39 in 2000 and 21 in 2001. Perhaps the Clinton administration directed more monitoring than did the George W. Bush administration. In contrast, the number of state inspections shows no obvious trend. The total number rises from a low of 594 in 1998 to crest at 647 in 2001, then falls back to 595 by 2002. In between, state inspections total 636 and 614 in 1999 and 2000, respectively.

In order to appreciate a more detailed trend in inspection use, Figure 7.1 displays the number of inspections performed in each month for the extended sample: January 1998 to March 2003. In particular, it also displays the monthly average number of inspections performed per facility that was active in a given month. The graph reveals that state inspections vary greatly from one month to next yet no obvious trend over time exists. Federal inspections also show no obvious trend over time. Any meaningful monthly variation in federal inspections is difficult to discern since federal inspections are frequently absent in a given month.

The use of inspections also varies across chemical manufacturing subsectors. The distribution is based on the set of survey respondents. At the 28 respondent facilities manufacturing inorganic chemicals, EPA performed 15 inspections. Similarly, EPA performed 17 inspections at the 52 respondent facilities manufacturing organic chemicals. In contrast, EPA performed only 2 inspections at the 17 respondent facilities manufacturing other

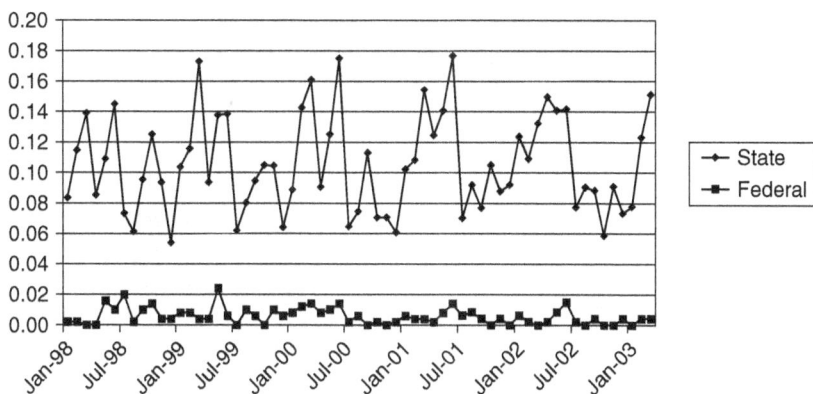

FIGURE 7.1

Average number of inspections per month per facility: State inspections and federal inspections

SOURCE: Environmental Protection Agency Permit Compliance System database.

chemicals. In comparison, state environmental agencies performed 202 inspections, 438 inspections, and 110 inspections, respectively, at inorganic chemical facilities, organic chemical facilities, and other chemical facilities. Relative to EPA regional offices, state agencies devoted more attention to organic chemical facilities and other chemical facilities.

Lastly, the use of inspections varies between smaller facilities and larger facilities, as measured by the number of workers employed at the individual facility. Using the median number of facility-level employees—266—to split the sample, 50 facilities are deemed smaller and 47 facilities are deemed larger. (Two respondent facilities do not provide data on the number of their employees; they are assumed to employ the mean number of employees, 342.) At the 50 smaller facilities, EPA and state agencies performed 21 and 327 inspections, respectively. In contrast, EPA and state agencies performed 13 and 423 inspections, respectively, at the 47 larger facilities. It appears that EPA devotes more attention to smaller facilities, while state agencies devote more attention to larger facilities.

7.6.3. Use of Inspections Across Time for Particular Facilities

Next we describe the use of inspections across time for particular facilities in order to explore the heterogeneous treatment of individual facilities within the set of survey respondents. For this exploration, we assess only those months in which a particular respondent facility is deemed active within the NPDES system. We also exclude inspections performed in 2003 so that the data are comparably populated across months, which avoids the concern of seasonal variation in inspections.

We first analyze the number of inspections performed per month against an individual facility. The average number of EPA inspections per month equals 0.0063. This average roughly implies a 0.6 percent chance of being inspected by EPA in any given month. In contrast, the average number of state inspections per month equals 0.132. This average roughly implies a 13 percent chance of being inspected by a state agency in any given month. These interpretations of the average values are only rough because a single facility may be inspected more than once in a given month. In the case of EPA inspections, facilities are not inspected in 99.5 percent of the months in the sample period and are inspected once in 0.46 percent of the months. However, facilities are inspected by EPA regional offices twice in three months and thrice in one month. The distribution of monthly state inspections also reveals multiple inspections in a given month. Facilities are not inspected by state agencies in 88 percent of the months in the sample period and are inspected once in 10.5 percent of the months. However, facilities are inspected by state agencies twice in 61 months, thrice in 6 months, and even 5 times in a single month.

To explore this point further, we assess the number of inspections performed at an individual facility over the course of a given calendar year. (We adjust for each facility's number of active months in a given year by inflating the number of annual inspections whenever the number of active months lies below 12.) On average, EPA performs 0.096 inspections per year and state agencies perform 2.06 inspections per year. Over a given year, 6 percent of the facilities are inspected at least once by EPA. In contrast, over a given year, 75 percent of the facilities are inspected at least once by state agencies.

Finally, we assess the number of inspections performed at an individual facility over the truncated sample period—January 1998 to December 2002—reported on an annualized basis. (Again, we adjust for each facility's number of active months.) The average facility is inspected 0.11 times per year by EPA and inspected 1.96 times per year by a state agency. Over the entire five-year period, 23 percent of the facilities are inspected at least once by EPA. In contrast, 97 percent of the facilities are inspected at least once by a state agency.

7.6.4. Distinguish by Type of Inspection

Our assessment of inspections up to this point distinguishes only between EPA inspections and state inspections. Thus, we treat each type of inspection performed by EPA or a state agency as fully comparable. However, both EPA and state agencies perform a variety of inspection types. EPA's PCS database identifies the following inspection types:

1. Performance audit inspections
2. Biomonitoring inspections
3. Compliance evaluation inspections
4. Reconnaissance inspections
5. Toxic inspections
6. Sampling inspections
7. Other inspections[11]

Helland (1998b) describes these inspection types. Reconnaissance inspections usually involve checking paperwork and are unlikely to detect noncompliant discharge levels. Reconnaissance inspections are generally used as fact-finding inspections and are not usually designed to detect noncompliance. Rather than identifying noncompliance, reconnaissance inspections serve to raise the regulatory presence of environmental protection agencies. These inspections are considered the least resource intensive of all inspections, especially since inspectors can complete these inspections within a few hours. Performance audit inspections and compliance evaluation inspections are designed to assess the facility's pollution-control equipment and paperwork with a focus on the facility's performance and compliance, as indicated by the category titles. However, neither type of inspection takes independent readings of discharges since neither samples a facility's discharge stream. These inspection types are considered more resource intensive than the reconnaissance inspections, especially since inspectors need more than a few hours, but generally less than a day, to complete them. In contrast to the aforementioned inspection types, the last three types of inspections sample a facility's discharge stream. The most common of these—sampling inspections—checks pollution-control equipment and paperwork and conducts an independent assessment of the facility's discharge. Compliance biomonitoring inspections use acute-toxicity testing to assess the sampled facility's discharge. Toxic inspections focus on toxic pollutants when assessing the sampled facility's discharge. These last three inspection types are the most resource intensive.

Based on the full set of Clean Water Act–regulated major facilities, EPA mostly uses compliance evaluation inspections (97). It never uses toxic inspections. In between, EPA performs 19 biomonitoring inspections, 18 audit inspections, 8 reconnaissance inspections, and 13 other inspections. Of particular interest, EPA performs only 20 sampling inspections, in contrast to the 155 nonsampling inspections. Similar to EPA, state agencies mostly use compliance evaluation inspections (1,300). However, in contrast to EPA, state agencies heavily use reconnaissance inspections (959), which represents the second-most frequently used type. At the low end, state agencies infrequently perform toxic inspections (60), similar to EPA, and other inspections (79). In between, state agencies perform 90 biomonitoring inspections and 82 audit inspections. Of particular interest, state agencies perform 654 sampling inspections, in contrast to the 2,571 nonsampling inspections. Thus, state agencies perform sampling inspections more frequently than do EPA regional offices as a proportion of overall inspections performed.[12]

7.7. USE OF ENFORCEMENT ACTIONS

We next describe the use of enforcement actions. Similar to our description of inspections, we explore enforcement actions taken against all Clean Water Act–regulated major facilities operating in the chemical manufacturing sector during our sample period and explore enforcement actions taken against only the surveyed facilities. As with inspections, when depicting use of enforcement in the full set of Clean Water Act–regulated chemical manufacturing facilities, we divide the sample into subsamples according to location (that is, EPA region) and time (for example, calendar year). When depicting use of enforcement in the set of surveyed facilities, we divide the sample into subsamples according to the chemical manufacturing subsector and facility size. Again we expand our sample window to include the year 1998. However, due to data availability, we must assess SEPs and injunctive relief sanctions over a truncated sample period: January 1998 to June 2001. All monetary values are adjusted for inflation, using 2000 as the base year.

7.7.1. Use of Enforcement for Entire Sample

We initially describe enforcement against the full set of Clean Water Act–regulated major chemical manufacturing facilities and then against the surveyed facilities. During the expanded sample period between January 1998 and March 2003, against the 536 Clean Water Act–regulated major facilities, EPA regional offices imposed 562 informal enforcement actions, EPA administrative courts imposed 251 formal nonsanctions and 9 fines totaling $303,355, and civil courts imposed 10 formal nonsanctions. Civil courts imposed no fines. Over the truncated sample period between January 1998 and June 2001, administrative courts imposed 6 injunctive relief sanctions totaling $15,277,773 and imposed no SEPs, while civil courts imposed no injunctive relief sanctions and imposed 1 SEP totaling $303,089. Of the 6 administrative injunctions, 3 are rather large—$1.3 million, $5.6 million, and $8.2 million—yet 2 seem trivially small: $914 and $3,565. Regardless, the use of nonfine sanctions clearly dominates the use of fines when measured in monetary terms.

Against the 97 survey respondents, EPA regional offices imposed 107 informal enforcement actions, EPA administrative courts imposed 54 formal nonsanctions and 1 fine, and civil courts imposed fewer than 10 formal nonsanctions. (We do not disclose the dollar value of the administrative fine nor the number of civil formal nonsanctions lest we identify respondent facilities.) As noted previously, civil courts imposed no fines against any major chemical manufacturing facility. Over the truncated sample period between January 1998 and June 2001, neither administrative courts nor civil courts imposed SEPs; administrative courts imposed a single injunctive relief sanction, while civil courts imposed no such sanctions. (We do not disclose the dollar value of the administrative injunction lest we identify the respondent facility.) For the sample of survey respondents, the use of fines is more prevalent than the use of SEPs and is comparable to the use of injunctions; when measured in monetary terms, the use of fines strongly dominates the use of injunctions.

The availability of data on SEPs and injunctive relief sanctions and the severely limited use of these sanction types, especially against survey respondent facilities, constrains our ability to examine the influences of SEPs and injunctions on environmental behavior and environmental performance. In particular, we are not able to examine the influence of SEPs imposed on a specific facility within our set of surveyed facilities because no SEPs were imposed. Moreover, our ability to examine the influence of injunctive relief sanctions imposed on a specific facility within our set of surveyed facilities relies on a single administrative sanction. As for the threat of SEPs based on the experience of all facilities within the sector of chemical manufacturing facilities, our analysis must rely upon a single civil SEP. Only the threat of injunctive relief sanctions remains a viable component to examine. Yet even this component is limited to administrative injunctions. For all these reasons, we scale back our analysis in Chapters 8 and 9 designed to capture the influences of SEPs and injunctive relief sanctions on environmental behavior and environmental performance, respectively.

7.7.2. Use of Enforcement Across Various Subsamples

Next we describe how the use of enforcement varies across EPA regions, time, subsectors within the chemical manufacturing sector, and different facilities based on size. First, enforcement varies across EPA regions. Based on the broad set of 536 Clean Water Act–regulated major facilities, at the 133 major facilities located within Region 4, EPA regional office imposed no informal enforcement actions and neither EPA administrative courts nor civil courts imposed any formal enforcement actions (including SEPs and injunctive relief sanctions). To be clear, facilities located in Kentucky, Tennessee, Mississippi, Alabama, Georgia, Florida, North Carolina, and South Carolina faced no federal enforcement at all. Against the 171 major facilities located within Region 6, EPA regional offices imposed 491 informal enforcement actions. EPA administrative courts imposed 249 formal nonsanctions, 3 penalties totaling $300,850, and 5 injunctive relief sanctions totaling a whopping $15,276,859. Civil courts imposed 8 formal nonsanctions, 1 SEP of $303,089, but no fines. Against the 232 major facilities located within the other EPA regions, EPA regional offices imposed 71 informal enforcement actions. EPA administrative courts imposed only 2 formal nonsanctions, 6 fines totaling only a trivial $2,505, 1 injunctive relief sanction of a mere $914, and no SEPs. Civil courts imposed 2 formal nonsanctions but neither fines, nor SEPs, nor injunctions.

The use of enforcement also varies over time between 1998 and 2002. Based on the broad set of 536 Clean Water Act–regulated major facilities, the number of informal enforcement actions drops from a high of 193 in 1998 to a low of 44 in 2001, with a slight bump to 60 in 2002. Thus, the use of enforcement declined over time under the Clinton administration and into the first year of the George W. Bush administration. The use of administrative formal nonsanctions rises from 79 in 1998 to 105 in 1999 but then drops to 17 by 2001, with a slight bump to 22 in 2002. The use of civil formal nonsanctions is nonexistent in 1998, 1999, and 2001 and barely existent in 2002 (1 action); in the exceptional year of 2000, civil courts imposed 9 formal nonsanctions. Administrative fines are nonexistent in 1998 and 2001 and barely existent in 2002 (1 fine of $1,245). During each of 1999 and 2000, administrative courts imposed 4 fines. However, the monetary total in 1999 is trivially small, $10; in contrast, the monetary total in 2000 equals $302,100. Thus, the use of administrative fines is trivial at best, except in 2000. Civil courts imposed no fines. Similarly, administrative courts imposed no SEPs. The single civil SEP is imposed in 1999. Administrative courts imposed 2 injunctions in 1998 totaling $5,721,364, 3 injunctions in 1999 totaling $8,230,501, and 1 injunction of $1,325,908 in 2000. Civil courts imposed no injunctions.

The use of enforcement also varies across chemical manufacturing subsectors. The distribution is based on the set of survey respondents. EPA regional offices imposed 30 informal enforcement actions against the 28 respondent facilities manufacturing inorganic chemicals, 73 informal actions against the 52 respondent facilities manufacturing organic chemicals, and only 4 informal actions against the 17 respondent facilities manufacturing other chemicals. The use of formal actions also varies across subsectors. Administrative courts imposed 14, 40, and 0 formal nonsanctions against respondent facilities manufacturing inorganic, organic, and other chemicals, respectively. All civil formal nonsanctions imposed against respondent facilities are targeted at facilities manufacturing organic chemicals. We do not report the distribution of fines, SEPS, and injunctions lest we identify individual survey respondents.

Lastly, the use of enforcement varies between smaller facilities and larger facilities. EPA regional offices imposed 85 informal enforcement actions against the 50 smaller facilities yet only 22 informal actions against the 47 larger facilities. Similarly, administrative courts imposed 48 formal nonsanctions against smaller facilities and only 6 against larger facilities. In contrast, civil courts imposed 84 percent of their formal nonsanctions against larger facilities. Again, we do not report the distribution of fines, SEPS, and injunctions lest we identify individual survey respondents.

7.7.3. Use of Enforcement Across Time for Particular Facilities

Next we describe the use of enforcement across time for particular facilities in order to explore the heterogeneous treatment of individual facilities within the set of survey respondents. For this exploration, we assess only those months in which a given respondent facility is deemed active within the NPDES system. We also exclude enforcement actions taken in 2003 so that the data are comparably populated across months, which avoids the concern of seasonal variation in enforcement. For this exploration, we focus exclusively on informal enforcement actions and formal nonsanctions since courts imposed only 1 fine, only 1 injunction, and no SEP against the survey respondents.

We first analyze the number of enforcement actions taken per month against an individual facility. The average number of informal enforcement actions per month equals 0.019. The average number of administrative formal nonsanctions per month equals 0.010; the civil counterpart equals 0.0015. In contrast to inspections, these average values do not "roughly" capture the likelihood of a single facility being enforced against in any given month because a single facility is generally enforced against more than once in a given month. In the case of informal actions, facilities are not enforced against in 98.6 percent of the months in the sample period and enforced against once in 0.8 percent of the months (41 months). However, facilities are enforced against multiple times in 19 months of the sample period, nearly half of the frequency with which facilities are enforced against only once in a given month. The distribution of formal nonsanctions also reveals multiple actions in a given month. Facilities are not formally enforced by administrative courts in 99.9 percent of the months in the sample period and are formally enforced once in a single month. However, facilities are formally enforced by administrative courts multiple times in 6 months of the sample period. Facilities are formally enforced by civil courts in 2 months: 1 month includes a single formal nonsanction and 1 month includes multiple formal nonsanctions.

To explore this point further, we assess the number of enforcement actions taken against an individual facility over the course of a given calendar year. (We adjust for each facility's number of active months in a given year by inflating the number of annual enforcement actions whenever the number of active months lies below 12.) On average, EPA regional offices imposed 0.22 informal enforcement actions per year. Over a given year, 7.2 percent of the facilities are informally enforced against at least once. Courts imposed formal nonsanctions less frequently. On average, administrative courts and civil courts imposed 0.114 and 0.017 formal nonsanctions per year, respectively. Over a given year, 1.5 percent and 0.4 percent of the facilities are formally enforced against at least once by administrative courts and civil courts, respectively.

Finally, we assess the number of enforcement actions taken against an individual facility over the truncated sample period—January 1998 to December 2002—reported on an annualized basis. (Again, we adjust for each facility's number of active months.) EPA regional offices imposed 0.219 informal actions per year against the average facility. Over the entire five-year period, 13.5 percent of the facilities are informally enforced against. Again, courts imposed formal nonsanctions less frequently. Administrative courts and civil courts, respectively, imposed 0.113 and 0.017 formal nonsanctions per year against the average facility. Over the entire five-year period, 5.2 percent and 2.1 percent of the facilities are formally enforced against at least once by administrative courts and civil courts, respectively, using nonsanctions.

7.8. LINKS FROM PERFORMANCE TO INTERVENTIONS: INSPECTIONS AND ENFORCEMENT ACTIONS

Lastly, we acknowledge that a link should connect poor environmental performance to decisions to perform inspections on the part of EPA regional offices and state environmental agencies and an even stronger link should connect noncompliance to decisions to take enforcement actions on the part of EPA regional offices, EPA administrative courts, and civil courts. Nevertheless, our study does not explore any of these links. Given the construction of our analysis in Chapters 8 and 9, the omission of these links from our analysis should not disrupt our conclusions to any meaningful degree. Any reader interested in analysis linking poor environmental performance, especially noncompliance, to inspection and enforcement decisions may consider analyses by Deily and Gray (1991), Gray and Deily (1996), Helland (1998a, 1998b), Earnhart (2004a, 2004c), Laplante and Rilstone (1996), and Nadeau (1997).

7.9. SUMMARY

Because discharge limits are meaningful only if there is a credible threat that they will be enforced, this chapter assesses what federal and state regulators are doing to induce compliance with discharge limits. In particular, it surveys what inspections and enforcement actions (which we refer to generically as government interventions) were taken by federal and state regulators against chemical manufacturing facilities during our sample period. After describing the legal authority to enforce discharge limits under the Clean Water Act, and the different types of inspections and enforcement actions available to regulators, we describe the two databases—PCS and EPA Docket—on which we drew information about inspections and enforcement actions. We also describe the limits of those databases. The chapter explores inspections and enforcement actions taken against all major facilities in the chemical industry regulated under the Clean Water Act during our sample period, as well as inspections and enforcement actions taken against only the facilities that participated in our survey. We also describe the use of inspections and enforcement actions across time for particular facilities in order to explore the heterogeneous treatment of individual facilities within the set of survey respondents. Our analysis reveals much about the kinds of inspections and enforcement actions taken (such as the fact that EPA mostly uses compliance evaluation inspections) and about how enforcement activity varies across EPA regions, time, subsectors within the chemical manufacturing sector, and size of facilities.

Chapter Seven Appendix

This appendix provides an assessment of inspection use and enforcement use in individual regions for certain years.

Chapter 8 utilizes the described data on inspections and enforcement to capture separately the threat of inspections and the threat of enforcement based on a notion of general deterrence. When constructing measures of general deterrence, our analysis calculates the use of inspections and enforcement in an individual region for a certain year. In order to represent more accurately the likelihood (that is, the threat) of an inspection or enforcement action, when constructing the general deterrence measures, our analysis controls for the number of active facilities per region by calendar year. In this chapter, we go one step further and control for the number of active months (summed across all facilities) per region by calendar year. Specifically, we count the inspections performed at or the enforcement actions taken against all facilities located in a given region and then divide by the number of active months experienced in the region in order to calculate the average number of inspections or enforcement actions per active month. Lastly, we inflate this average by 12 in order to annualize the region-year-specific measures. For comparability with the general deterrence measures constructed in Chapter 8, our analysis in this chapter explores inspections performed at and enforcement actions taken against all 536 Clean Water Act–regulated major facilities.

The resulting measures indicate that the threat of both EPA inspections and state inspections and the threat of both informal and formal enforcement vary across regions and across time. Consider the average number of EPA inspections per facility per year. This average varies between 0.008 (Region 4 in 2002) and 0.145 (Region 6 in 1999). Even within a region, the average number of EPA inspections varies. For example, in Region 6, the average number of EPA inspections varies between 0.019 (in 2002) and 0.145 (in 2002). In addition, the average number of EPA inspections varies within a year. For example, in 1998, the average number of EPA inspections varies between 0.047 (in Region 4) and 0.123 (in other regions). Regarding enforcement, consider the average number of administrative formal nonsanctions per facility per year. This average varies between 0 (all years in Region 4 and most years in other regions) and 0.663 (Region 6 in 1999). The average number of administrative formal nonsanctions per facility never varies from 0 in Region 4. In contrast, the average number varies in Region 6 substantially between 0.106 (in 2001) and 0.663 (in 1999). In other regions, the average number of administrative formal nonsanctions is 0 for all years except 2002, when it equals 0.011. For the interested reader, a full tabular display is provided on the Web site of the Stanford University Press.

In the case of inspections, adjusting for the number of active facilities or number of active months in a given region during a particular year seems reasonable since any facility faces the threat of an inspection. As noted in Section 7.3, EPA guidelines prompt EPA regional offices and state environmental agencies to inspect each major facility annually. However, in the case of enforcement, adjusting for the number of active facilities or active months may not seem reasonable since enforcement is presumably aimed at noncompliant performance, which may vary across regions and over time. Thus, rather than adjusting for variation in the number of active facilities or months, the analysis should adjust for the number of noncompliant facilities or number of noncompliant months per region by year. This improvement is certainly beneficial but challenging to implement since noncompliance does not immediately prompt enforcement and multiple months of noncompliance may prompt enforcement.

As an alternative strategy, we assess whether the degree of noncompliance differs across regions and over time. For this assessment, we separately calculate the number of noncompliant facilities and the number of noncompliant months in a given region during a particular year as a ratio of the relevant number of active facilities or active months. We do not limit the analysis to noncompliance relating to only total suspended solids (TSS) and biological oxygen demand (BOD) discharge limits. Instead, we consider noncompliance relating to all permitted discharge limits. In order to avoid concerns about seasonal variation in compliance, we ignore the first three months of 2003.

The resulting ratios of noncompliant to active facilities or months reveal some variation across regions (Region 4, Region 6, other regions) and over time. Consider first the ratio of noncompliant facilities to active facilities. We initially assess variation over time for any given region. For Region 4, the ratio varies between 53 percent (in 2002) and 64 percent (in 1998) over time. For Region 6, the ratio varies between 56 percent (in

2002) and 68 percent (in 1998) over time. For other regions, the ratio varies between 59 percent (in 2002) and 65 percent (in 2000). In all three cases, the variation is meaningful but not dramatic. We next assess variation across regions for a given year. In 1998, the ratio varies between 62 percent and 68 percent. In 1999, the ratio varies between 60 percent and 68 percent. In 2000, the ratio varies between 64 percent and 67 percent. In 2001, the ratio varies between 53 percent and 62 percent. In 2002, the ratio varies between 53 percent and 59 percent. Again, in all five cases, the variation is meaningful but not dramatic.

An assessment of the ratios of noncompliant to active months reveals a similar degree of variation across regions and over time. The results of our two assessments compel us to acknowledge that our failure to adjust for noncompliance across regions and over time may distort our understanding of the threat of enforcement. Still, based on our assessments, we believe that the degree of distortion is manageable.

CHAPTER EIGHT

Effect of Government Interventions
on Environmental Behavior

This chapter explores the effect of government interventions on environmental behavior. As the key purpose of this exploration, we seek the answer to the following research question: do government interventions help to induce better behavior? To answer this policy-relevant research question, we attempt to link the use of government interventions to facilities' behavioral decisions. To establish this link, we must first describe deterrence and then distinguish between specific deterrence and general deterrence. With this description and distinction in hand, we employ two empirical approaches to generate evidence that helps to answer our research question.

8.1. DETERRENCE: SPECIFIC DETERRENCE
AND GENERAL DETERRENCE

To establish the link from government interventions to environmental behavioral decisions, we first describe deterrence and then distinguish between specific deterrence and general deterrence.

Deterrence stems from inspections and enforcement. First, the frequency and intensity of surveillance through agency inspections would appear to influence the perceptions of regulated entities as to the probability of detection and punishment. Accordingly, an increase in the frequency of inspections should cause inspection targets to raise their estimates of the risks of detection and punishment and to react by taking steps to reduce the incidence of noncompliance with applicable environmental regulations and permit provisions. In addition to increasing the likelihood of an eventual sanction, inspections may impose costs on the facilities being inspected by disrupting production or requiring employees to spend time accompanying inspectors or generating information for them instead of fulfilling their normal responsibilities. One therefore would expect regulated facilities to improve environmental performance in response to inspections they have experienced or the threat of inspections.

As important, deterrence stems from enforcement. It is EPA's position that enforcement actions serve to engender better environmental behavior. According to EPA (1998, pp. 24, 796–97), for example, "Penalties promote environmental compliance and help protect public health by deterring future violations by the same violator and deterring violations by other members of the regulated community"; they "help ensure a national level playing field by ensuring that violators do not obtain an unfair economic advantage over their competitors who made the necessary expenditures to comply on time"; and they "encourage regulated entities to adopt pollution prevention and recycling techniques in order to minimize their pollutant discharges and reduce their potential liabilities."

Other forms of enforcement can have the same kinds of deterrent effects as monetary fines. The use of injunctions has the capacity to create a general deterrent effect if a regulated entity perceives their use as creating the threat of the issuance of an injunction in the future against that facility. The imposition on regulated facilities of the obligation to perform supplemental environmental projects (SEPs), which may be costly, can serve the same function. As indicated in Chapter 7, however, SEPs may actually lower the cost to regulated entities of regulatory violations, which may create underdeterrence of regulatory violations for the first time or exacerbate previously existing underdeterrence (Dana, 1998, p. 1184). Informal actions are designed largely to notify a facility that regulators believe that a violation is occurring and to prompt action by the facility to eliminate the violation. EPA defines an informal action as "telephone calls, a letter, or issuing a notice of violation, all of which simply communicates to the permittee that a violation was detected and should be corrected (EPA, 1990, 6-3). Those kinds of informal actions are also potential specific deterrents, as they clearly communicate to the facility concerned that regulators have gathered information that they regard as sufficient to prove a violation and are prepared to pursue enforcement action if the violator does not cure the problem. Finally, the formal nonsanction is analogous to a judicially issued injunction in terms of its deterrent impact. A formal nonsanction presumably includes formal enforcement action other than a monetary fine. It therefore would include an administrative injunction (sometimes referred to as an administrative compliance order). Because administrative injunctions and judicial injunctions have the same legal impact (that is, they mandate compliance and set a legally enforceable schedule for doing so), they ought to have the same deterrent effects. It is possible that administrative injunctions have a less weighty general deterrent impact, however, if their issuance is less widely publicized than the issuance of judicial injunctions.

We next distinguish between specific deterrence and general deterrence.

8.1.1. Specific Deterrence

One form of deterrence—specific deterrence—stems from interventions against a particular facility in the recent past. Accordingly, the analysis uses lagged values of interventions to construct the factors designed to capture specific deterrence. In particular, we capture specific deterrence by calculating the cumulative count or dollar value of interventions performed against a specific facility in the twelve-month period preceding the identified environmental behavior. In the case of inspections, informal enforcement actions, and formal nonsanction enforcement actions, we calculate the cumulative count of interventions performed. In the case of federal sanctions, such as fines, we calculate the cumulative dollar value of federal sanctions imposed against a specific facility in the preceding twelve-month period.

Clearly, we could construct specific deterrence measures in a variety of ways, so we should explain our chosen approach. First, we choose a period of twelve months for two reasons: (1) major polluters should be inspected once per year, and (2) previous studies, such as Laplante and Rilstone (1996) and Earnhart (2004b, 2004c), examine a twelve-month period of lagged interventions. Second, the chosen approach of accumulating interventions is more consistent with reality than the alternative approach of including multiple monthly measures of lagged interventions (for example, Magat and Viscusi, 1990). According to EPA officials, regulatory agencies generally induce better behavior by repeatedly inspecting polluters. The same logic may hold for enforcement sanctions. Third, the chosen approach of cumulative interventions retains the explanatory power of potentially multiple interventions within a single deterrence factor rather than dissipating the explanatory power across several factors.

This depiction applies to our construction of specific deterrence measures in general. However, we must formulate the data on government interventions so they match with our data on behavior. Recall that some behavioral forms are measured for each of three calendar years between

1999 and 2001, inclusively. Other behavioral forms are measured for the twelve-month period preceding the completion of the survey. The remaining behavioral forms are measured for the three-year period preceding the completion of the survey. Since the time frame differs across three broad categories of behavioral forms, our construction of specific deterrence measures also differs across the three time frames. For the behavioral forms measured in each of three calendar years, the specific deterrence measures include the government interventions against a particular facility over the twelve-month period preceding each calendar year. For the behavioral forms measured in the twelve-month period or three-year period preceding survey completion, the specific deterrence measures include the government interventions against a particular facility over the twelve-month period preceding each facility's particular twelve-month period or three-year period of behavior, respectively.

8.1.2. General Deterrence

The other form of deterrence—general deterrence—stems from the threat of interventions against facilities in general within the near future based on the experience of other related facilities. The notion of general deterrence assumes that regulated facilities possess some knowledge of the inspections and enforcement actions taken against other regulated facilities. That assumption may be called into question in particular contexts for particular facilities, especially if regulators do not take steps to publicize the interventions they conduct. Our treatment of general deterrence is consistent with the general treatment of that concept within the environmental enforcement literature by assuming at least some such knowledge on the part of regulated facilities regarding monitoring and enforcement efforts directed at other facilities.

To generate the general deterrence measures, we calculate the annual aggregate measures of government interventions against *other* similar facilities: major chemical facilities in the same relevant location (for example, state) and same time period (Earnhart, 2004c; Nadeau, 1997). In the case of inspections, informal enforcement actions, and formal non-sanction enforcement actions, we calculate the aggregate count of interventions performed against other similar facilities. In the case of federal sanctions, we calculate the aggregate dollar value of federal sanctions imposed against other similar facilities. We calculate the aggregate count of state inspections by state and calculate the aggregate count or value of each federal intervention by EPA region. To adjust for differences in the number of major chemical facilities across states or EPA regions and across time, the analysis divides each aggregate measure of government interventions by the number of other major chemical facilities in each state or EPA region of the given year.

Clearly, we could construct general deterrence measures in a variety of ways, so we should discuss the implications of our chosen approach. First, similar to most previous studies of inspections (for example, Earnhart, 2004c; Laplante and Rilstone, 1996; Gray and Deily, 1996; Nadeau, 1997), our analysis focuses exclusively on the likelihood of an inspection when capturing the threat of an inspection. Second, by distinguishing across informal enforcement actions, formal nonsanctions, and sanctions, our analysis at least crudely controls for the burden imposed by an enforcement action when capturing the threat of enforcement. Third, when examining the threat of sanction-based enforcement, our analysis jointly considers the likelihood of a sanction and the burden imposed by any sanction by examining the aggregate value of sanctions (in dollars) imposed against other similar facilities. (By dividing this aggregate value by the number of other similar facilities, the resulting measure represents the unconditional average sanction amount imposed against other similar facilities.)

This depiction applies to our construction of general deterrence measures in general. However, we must formulate the data on government interventions so they match our data on behavior. As with specific deterrence measures, our construction of general deterrence measures differs across the time frames for the three broad categories of behavioral forms. For the behavioral forms measured in each of three calendar years, the general deterrence measures include the government interventions taken against other facilities over the calendar year on a per-facility basis. For the behavioral forms measured in the twelve-month period preceding survey completion, the general deterrence measures include the government interventions against other facilities over the relevant calendar year on a per-facility basis. For those facilities completing the survey in 2002, we use 2002 as the relevant calendar year; the same logic applies to facilities completing the survey in 2003. For the behavioral forms measured in the preceding three-year period, the general deterrence measures include the government interventions against other facilities over each relevant calendar year on a per-facility basis, averaged across the relevant calendar years. For those facilities completing the survey in 2002, the relevant calendar years are 2000, 2001, and 2002; for those facilities completing the survey in 2003, the relevant calendar years are 2001, 2002, and 2003.

Lastly, we acknowledge that our measure of general deterrence may be flawed. If individual facilities believe that monitoring and enforcement resources are fixed, then an increase in one of our aggregate-based general deterrence measures may reflect a *decrease* in the perceived threat of an intervention against the individual facility. Fortunately, this possibility is reduced by our consideration of interventions against other facilities over an entire calendar year. In this way, the perceived threat is based on both the recent (past) use of government interventions and the future use of government interventions against others. While the former basis is vulnerable to the noted concern since the recent interventions clearly were not performed against the individual facility, the latter basis is not vulnerable, reflecting instead the imminent threat of interventions, which still may be performed against the individual facility.

8.2. EMPIRICAL APPROACHES AND EVIDENCE

With the distinction between specific deterrence and general deterrence established, we employ two empirical approaches that attempt to link the use of government interventions to facilities' behavioral decisions, in the process generating evidence that helps to assess whether government interventions help to induce better behavior. The first empirical approach splits the sample of facilities into two subsamples based on the degree of deterrence faced by an individual facility and then explores whether behavior is better when deterrence is greater. As a complement to this approach, we assess the correlation between deterrence and behavioral types that are recorded as quantitative measures (for example, external consulting expenditures). The second empirical approach employs multivariate regression analysis, which uses particular techniques, such as ordinary least squares regression, to estimate a functional relationship between a specific outcome (that is, a single behavioral decision) and various explanatory factors, with a focus on the factors relating to government interventions.

8.2.1. Split Sample Analysis

8.2.1.1. Formulation of Hypothesis; Framework for Testing Hypothesis. To repeat, the first empirical approach splits the sample of facilities into two subsamples based on the degree of deterrence faced by an individual facility, as identified by the median degree of deterrence, and then explores whether behavior is better when deterrence is greater by employing two-sample tests. As our general hypothesis, we test whether greater deterrence prompts better environmental behavior. Based on our testing, we interpret our answer within the following light: if behavior improves as deterrence increases, then government interventions would appear effective at prompting better behavior.

In order to test this general hypothesis, we examine the effects of government intervention–based deterrence on the various measures of environmental behavior. For this examination, we discern between qualitative measures of behavior, which are identified by categories (for example, presence versus absence of ISO 14001 compliance), and quantitative measures of behavior, which are identified by a number (for example, number of environmental employees). The former set of measures includes both qualitative measures that include only two categories (for example, ISO 14000 compliance: presence versus absence) and qualitative measures that include more than two categories (for example, audit team composition: none, only internal members, only external members, both internal and external members). As performed in Chapter 5, in order to interpret our analysis with more clarity, we reformulate each qualitative measure with more than two categories into a qualitative measure with only two broader categories.

We employ two analytical strategies. For all of the environmental behavior measures, we compare the distributions of behavioral measures for the two subsamples: facilities facing below-median deterrence and facilities facing above-median deterrence. Based on our hypothesis, facilities facing greater (above-median) deterrence should generate a higher distribution of environmental behavior than facilities facing lesser (below-median) deterrence. By higher distribution, we mean a distribution located at higher values of environmental behavior (for example, more audits conducted per year). For the quantitative behavioral measures, we also assess the correlation between the extent of environmental behavior measure and the degree of government intervention–based deterrence. According to our hypothesis, we expect a positive correlation between these two factors.

Readers less interested in the details of our statistical analysis than in the results of our statistical analysis may wish to skip over the following three paragraphs or even to the beginning of Section 8.2.1.2, in which we interpret the results of our hypothesis testing.

For qualitative measures of environmental behavior, we employ a two-sample test to assess whether any identified difference in the subsample-specific distributions is statistically significant. For the quantitative measures of environmental behavior, we also assess whether the distribution of a particular behavioral measure differs between the subsample of facilities facing below-median deterrence and the subsample of facilities facing above-median deterrence. In particular, we employ two two-sample tests—Wilcoxon two-sample test and median two-sample test—in order to assess whether any identified difference in the subsample-specific distributions is statistically significant. We report only the former test results because the latter test results generate identical conclusions (with the few exceptions noted).[1]

Two-sample tests assess whether the distribution of one subsample is located at higher levels of environmental behavior than the distribution of the other subsample. Any significant difference between the two distributions provides evidence for or against our hypothesis. We deem a difference between two distributions statistically significant when our analysis is able to reject the null hypothesis of zero difference at a significance level of 10 percent or less, as revealed by a p-value of 0.10 or less. We deem a difference between two distributions statistically insignificant when our analysis is not able to reject the null hypothesis of zero difference at a significance level of 10 percent or less, as revealed by a p-value greater than 0.10.

In addition to the two-sample tests, for the quantitative behavioral measures, we also assess the pairwise correlation between a behavioral measure and deterrence by calculating a Pearson correlation coefficient. We assess the signs of these correlations in order to explore the types of relationships between particular types of deterrence and particular measures of environmental behavior. A positive correlation indicates that the extent of environmental behavior and degree of deterrence generally vary together in the same direction. A negative correlation indicates the opposite. According to our hypothesis, we expect a positive correlation. We also assess the magnitude of these correlations in order to explore the strength of the relationships between deterrence and environmental behavior. In addition, we assess the statistical significance of these correlations, as captured by the p-values associated with the correlations, in order to test whether the calculated magnitudes are distinguishable from zero, which implies no relationship between an individual environmental performance measure and an individual discharge limit level. We deem a correlation as "statistically significant" when our analysis is able to reject the null hypothesis of a zero correlation at a significance level of 10 percent or less, as revealed by a p-value of 0.10 or less.

In the preceding chapter, we describe and explore eleven types of government interventions: federal inspections, state inspections, informal enforcement actions, administrative formal nonsanctions, civil formal nonsanctions, administrative fines, civil fines, administrative injunctive relief, civil injunctive relief, administrative SEP, and civil SEP. As noted earlier in this chapter, we explore two forms of deterrence: specific deterrence and general deterrence. In total, we explore the effects of twenty-two different deterrence measures on environmental behavioral decisions.

An exploration of the links between each of twenty-two deterrence measures and each of twenty-eight behavioral measures would be unwieldy.

Rather than exploring these numerous links using two-sample tests, we
save this exploration for the second empirical approach—multivariate
regression analysis, which is better able to address a variety of links. For
the empirical approach of two-sample testing, we combine federal inspections and state inspections into a single measure of inspection-related
deterrence. In the case of inspection-related specific deterrence, the single
measure represents the sum of federal and state inspections in the relevant
preceding twelve-month period. In the case of inspection-related general
deterrence, the single measure represents the average of the federal and
state inspection-related general deterrence measures. We are not able to
sum federal and state inspections because the denominators of the two
general deterrence measures differ since they reference different areas—
EPA regions versus states.

Similar to our treatment of inspections, we combine all enforcement
actions into a single measure of enforcement-related deterrence. Regardless of the form of deterrence—specific or general—we calculate the
count of all types of enforcement actions. Clearly, we are not able to retain the dollar values of sanctions because neither informal actions nor
formal nonsanction actions are measured in dollar terms. This count of
enforcement actions excludes injunctions and SEPs for reasons explained
immediately following.

As noted in the preceding chapter, our data on injunctive relief and SEPs
are limited to the period between 1998 and 2001. These data are sufficient
for our exploration of environmental behavior measured for each calendar year between 1999 and 2001. However, these data are insufficient for
our exploration of environmental behavior measured for either the twelve-
month period preceding or the three-year period preceding each facility's
survey completion. Thus, we limit our two-sample testing and correlation
assessment regarding injunctive relief and SEPs to environmental behavior measured for each calendar year between 1999 and 2001. Since our
analysis addresses injunctive relief and SEPs separate from other enforcement actions, we are able to retain dollar value as our unit of measurement.
In order to reduce the number of links being explored, we combine
administrative injunctions and civil injunctions into a single measure of
injunction-related deterrence and combine administrative SEPs and civil
SEPs into a single measure of SEP-related deterrence.[2]

8.2.1.2. Hypothesis Testing.   Given all of this framing, we explore the
effect of deterrence on various forms of environmental behavior. We assess first the influence of inspections, second the influence of enforcement
actions, and lastly the influence of injunctions and SEPs. In the end, we
attempt to draw a conclusion about the influence of deterrence, in gen-

eral, on each form of environmental behavior. In all cases, we attempt to identify evidence that either supports or rejects our overall hypothesis that greater deterrence induces facilities to employ stronger environmental behavior (for example, conduct more audits per year). Since the two-sample testing, in combination with the correlation assessment, represents the weaker of the two empirical approaches, we do not evaluate the two-sample tests and correlations in depth, saving attention for the stronger of the two empirical approaches: multivariate regression analysis.

8.2.1.2.1. Hypothesis Testing: Inspection-Related Deterrence. We first explore inspection-related deterrence. The two-sample test results indicate that the influence of inspection-related specific deterrence varies across the many forms of environmental behavior. Consistent with our hypothesis, inspection-related specific deterrence positively influences the number of environmental employees—in absolute terms and relative to overall employees—and audit count.[3] Inconsistent with our hypothesis, inspection-related specific deterrence negatively influences environmental employees' education and external consultant absolute expenditures. Providing no evidence for our hypothesis, inspection-related specific deterrence does not appear to influence the remaining forms of environmental behavior. Table 8.1.a fully tabulates the test results. All of the correlations appear to indicate that inspection-related specific deterrence does not influence (significantly) quantitative measures of environmental behavior.[4]

The two-sample test results indicate that, similar to inspection-related specific deterrence, the influence of inspection-related general deterrence varies across the many forms of environmental behavior. Consistent with our hypothesis, inspection-related general deterrence positively influences environmental employees relative to overall employees, external consultant expenditures relative to overall employees,[5] and management steps to communicate environmental performance goals and targets to facility employees. Inconsistent with our hypothesis, inspection-related general deterrence negatively influences environmental employees' work experience, training attendance, and audit count.[6] Providing no evidence for our hypothesis, inspection-related general deterrence does not appear to influence the remaining forms of environmental behavior. Table 8.1.b fully tabulates the test results.

All correlations but one indicate that inspection-related general deterrence does not influence (significantly) the quantitative measures of environmental behavior. As the single exception, inconsistent with our hypothesis, inspection-related general deterrence negatively influences environmental employees' education ($\rho = -0.175$, $p = 0.091$).

TABLE 8.1

Effect of deterrence on environmental behavioral decisions

8.1.a. Inspection-based specific deterrence

8.1.a.i. Qualitative measures of environmental behavior

Behavioral Measure	Category	Frequency (subsample conditional percentage)[a]		Test Statistic (p-value)[b]
		Below-Median-Deterrence Subsample	Above-Median-Deterrence Subsample	
Training: Quality	Lower: none, internal only, external only	36	20	0.329
	Higher: internal and external	29 (44.62)	12 (37.50)	(1.000)
TSS Treatment Presence	Absent	NS^c	NS^c	0.348
	Present	NS^c	NS^c	(1.000)
TSS Treatment Extent	Lesser: none, coagulation + sedimentation, sedimentation in effluent pond	32	14	0.221
	Greater: sedimentation in final clarifier, granular media filtration, membrane filtration	32 (50.00)	17 (54.84)	(1.000)
BOD Treatment Presence	Absent	12	6	0.000
	Present	52 (81.25)	26 (81.25)	(1.000)
BOD Treatment Extent	Lesser: none, solids removal, stabilization pond, trickling filter	30	11	0.577
	Greater: activated sludge, chemical oxidation, carbon adsorption	34 (53.13)	21 (65.63)	(0.893)

Variable	Category			Statistic (p-value)
Treatment Upgrade: Presence	No	19	12	0.706
	Yes	48 (71.64)	15 (55.56)	(0.702)
Treatment Upgrade: Designed to Reduce Wastewater	No	29	15	0.478
	Yes	36 (55.38)	12 (44.44)	(0.976)
Monitoring Program	Lesser: none (except final discharge), treatment process, single-point sewer and treatment process	27	21	1.083 (0.192)
	Greater: multiple-point sewer and treatment process	37 (57.81)	11 (34.38)	
Audit: Team Composition	Lesser: none, visiting corporate/consultants, teams of facility employees	24	10	0.400 (0.997)
	Greater: teams of visiting corporate/consultants and facility employees	36 (60.00)	22 (68.75)	
Audit: Classification Protocol	Lesser: none, noncompliance findings, noncompliance versus other, priority categories of vulnerability	47	23	0.019 (1.000)
	Greater: ranked serially by vulnerability for noncompliance	16 (25.40)	8 (25.81)	
External Consultants: Use	No	15	6	0.021
	Yes	54 (78.26)	21 (77.78)	(1.000)
Requested Assistance from Wastewater Regulator	No	50	21	0.187
	Yes	18 (26.47)	6 (22.22)	(1.000)
Compliance Assistance Meets Facility's Environmental Employees' Needs	Lesser: never, some of time, most of time	46	13	0.389 (1.000)
	Greater: always	21 (31.34)	9 (40.91)	

(continued)

TABLE 8.1 (*continued*)

8.1.a. Inspection-based specific deterrence

8.1.a.i. Qualitative measures of environmental behavior

Behavioral Measure	Category	Frequency (subsample conditional percentage)[a]		Test Statistic (p-value)[b]
		Below-Median-Deterrence Subsample	Above-Median-Deterrence Subsample	
Importance Management Places on Role of General Plant Employees in Identifying and Correcting Conditions Leading to Noncompliance	Lesser: somewhat important, important	11	5	0.072
	Greater: very important	53	27	(1.000)
		(82.81)	(84.38)	
ISO 14001 Compliance	No	56	23	0.359
	Yes	6	5	(1.000)
		(9.68)	(17.86)	

[a] The subsample conditional percentage is reported only for the category that represents either the presence of a behavioral measure or the greater extent of a behavioral measure. For example, in the subsample of facilities facing below-median deterrence, the reported percentage indicates the number of observations where the facility employs the identified behavior or the greater extent of behavior, while also facing below-median deterrence, relative to the number of all facilities facing below-median deterrence.

[b] Test statistic applies to the comparison between the below-median-deterrence subsample and the above-median-deterrence subsample.

[c] NS indicates that at least one cell in the cross-tabulation includes three or fewer observations if the cell is populated.

8.1.a.n. Quantitative measures of environmental behavior

Behavioral Measure	Deterrence Subsample[a]	N	Mean	Sign of Difference in Means[b]	Test Statistic (p-value)[c]
Environmental Employees: Absolute	Below median	190	96.355	+	1.857 (0.032)
	Above median	90	111.239		
Environmental Employees: Relative to Overall Employees	Below median	185	0.063	−	1.049 (0.147)
	Above median	88	0.035		
Wastewater Employees: Absolute	Below median	63	13.214	−	−0.253 (0.400)
	Above median	32	9.500		
Wastewater Employees: Relative to Overall Employees	Below median	61	0.090	+	−0.256 (0.399)
	Above median	31	0.128		
Environmental Employees' Education	Below median	65	59.446	−	−1.918 (0.028)
	Above median	29	43.931		
Environmental Employees' Work Experience	Below median	63	13.635	−	−0.449 (0.327)
	Above median	32	13.156		
Training Days	Below median	63	8.790	+	−0.318 (0.375)
	Above median	28	12.540		
Training Attendance	Below median	65	91.077	−	−0.497 (0.309)
	Above median	29	87.483		
Audit: Annual Count	Below median	173	6.116	−	1.448 (0.074)
	Above median	89	4.944		
External Consultants: Annual Expenditures—Absolute	Below median	64	39,828	−	−1.243 (0.107)
	Above median	24	16,771		
External Consultants: Annual Expenditures—Relative to Overall Employees	Below median	63	327.393	−	−0.612 (0.270)
	Above median	24	158.688		
Management's Steps to Communicate Environmental Performance Goals and Targets to Facility Employees (count)	Below median	65	3.262	+	0.420 (0.337)
	Above median	32	3.469		
Management's Steps to Communicate Environmental Performance Goals and Targets to External Parties (count)	Below median	65	2.338	−	−0.117 (0.453)
	Above median	32	2.281		

[a] Deterrence categories: below median deterrence, above median deterrence.

[b] The difference in mean values = (mean of the below-median-deterrence subsample) − (mean of the above-median-deterrence subsample).

[c] Test statistic applies to the comparison between the below-median-limit subsample and above-median-limit subsample; it reports the Wilcoxon two-sample test statistic in the form of a normal approximation; the associated p-value stems from a one-tailed test.

(continued)

TABLE 8.1 (*continued*)

8.1.b. Inspection-based general deterrence

8.1.b.i. Qualitative measures of environmental behavior

Behavioral Measure	Category	Frequency (subsample conditional percentage)[a]		Test Statistic (p-value)[b]
		Below-Median-Deterrence Subsample	Above-Median-Deterrence Subsample	
Training: Quality	Lower: none, internal only, external only	28	28	0.176
	Higher: internal and external	22	19	(1.000)
		(44.00)	(40.43)	
TSS Treatment Presence	Absent	5	6	0.138
	Present	44	40	(1.000)
		(89.80)	(86.96)	
TSS Treatment Extent	Lesser: none, coagulation+sedimentation, sedimentation in effluent pond	25	21	0.261
	Greater: sedimentation in final clarifier, granular media filtration, membrane filtration	24	25	(1.000)
		(48.98)	(54.35)	
BOD Treatment Presence	Absent	7	11	0.447
	Present	42	36	(0.988)
		(85.71)	(76.60)	
BOD Treatment Extent	Lesser: none, solids removal, stabilization pond, trickling filter	24	17	0.627
	Greater: activated sludge, chemical oxidation, carbon adsorption	25	30	(0.826)
		(51.02)	(63.83)	

Variable	Category			
Treatment Upgrade: Presence	No	18	13	0.380 (0.999)
	Yes	31 (63.27)	32 (71.11)	
Treatment Upgrade: Designed to Reduce Wastewater	No	22	22	0.200 (1.000)
	Yes	26 (54.17)	22 (50.00)	
Monitoring Program	Lesser: none (except final discharge), treatment process, single-point sewer and treatment process	22	26	0.510 (0.957)
	Greater: multiple-point sewer and treatment process	27 (55.10)	21 (44.68)	
Audit: Team Composition	Lesser: none, visiting corporate/consultants, teams of facility employees	17	17	0.077 (1.000)
	Greater: teams of visiting corporate/consultants and facility employees	30 (63.83)	28 (62.22)	
Audit: Classification Protocol	Lesser: none, noncompliance findings, noncompliance versus other, priority categories of vulnerability	37	33	0.413 (0.996)
	Greater: ranked serially by vulnerability for noncompliance	10 (21.28)	14 (29.79)	
External Consultants: Use	No	16	5	1.078 (0.195)
	Yes	33 (67.35)	42 (89.36)	
Requested Assistance from Wastewater Regulator	No	38	33	0.130 (1.000)
	Yes	12 (24.00)	12 (26.67)	
Compliance Assistance Meets Facility's Environmental Employees' Needs	Lesser: never, some of time, most of time	32	27	0.319 (1.000)
	Greater: always	14 (30.43)	16 (37.21)	

(continued)

TABLE 8.1 (*continued*)

8.1.b. Inspection-based general deterrence

8.1.b.i. Qualitative measures of environmental behavior

| Behavioral Measure | Category | Frequency (subsample conditional percentage)[a] | | Test Statistic (p-value)[b] |
		Below-Median-Deterrence Subsample	Above-Median-Deterrence Subsample	
Importance Management Places on Role of General Plant Employees in Identifying and Correcting Conditions leading to Noncompliance	Lesser: somewhat important, important	9	7	0.170
	Greater: very important	40	40	(1.000)
		(81.63)	(85.11)	
ISO 14001 Compliance	No	43	36	0.368
	Yes	4	7	(0.999)
		(8.51)	(16.28)	

[a] The subsample conditional percentage is reported only for the category that represents either the presence of a behavioral measure or the greater extent of a behavioral measure.

[b] Test statistic applies to the comparison between the below-median-deterrence subsample and the above-median-deterrence subsample.

8.1.b.ii. Quantitative measures of environmental behavior

Behavioral Measure	Deterrence Subsample[a]	N	Mean	Sign of Difference in Means[b]	Test Statistic (p-value)[c]
Environmental Employees: Absolute	Below median	144	99.302	+	0.373
	Above median	136	103.085		(0.355)
Environmental Employees: Relative to Overall Employees	Below median	140	0.073	–	0.297
	Above median	133	0.034		(0.383)
Wastewater Employees: Absolute	Below median	50	10.650	+	0.886
	Above median	45	13.422		(0.196)
Wastewater Employees: Relative to Overall Employees	Below median	48	0.117	–	1.063
	Above median	44	0.088		(0.144)
Environmental Employees' Education	Below median	48	57.833	–	-0.968
	Above median	46	51.348		(0.167)
Environmental Employees' Work Experience	Below median	50	14.130	–	-1.331
	Above median	45	12.744		(0.092)
Training Days	Below median	47	10.500	–	-0.451
	Above median	44	9.349		(0.326)
Training Attendance	Below median	49	88.714	+	-0.048
	Above median	45	91.333		(0.048)
Audit: Annual Count	Below median	130	5.808	–	-1.768
	Above median	132	5.629		(0.039)
External Consultants: Annual Expenditures—Absolute	Below median	45	29,244	+	1.167
	Above median	43	38,035		(0.122)
External Consultants: Annual Expenditures—Relative to Overall Employees	Below median	44	358.489	–	1.535
	Above median	43	201.412		(0.062)
Management's Steps to Communicate Environmental Performance Goals and Targets to Facility Employees (count)	Below median	50	3.160	+	1.296
	Above median	47	3.511		(0.097)
Management's Steps to Communicate Environmental Performance Goals and Targets to External Parties (count)	Below median	50	2.200	+	0.997
	Above median	47	2.447		(0.159)

[a]Deterrence categories: below median deterrence, above median deterrence.

[b]The difference in mean values = (mean of the above-median-deterrence subsample) – (mean of the below-median-deterrence subsample).

[c]Test statistic applies to the comparison between the below-median-limit subsample and above-median-limit subsample; it reports the Wilcoxon two-sample test statistic in the form of a normal approximation; the associated p-value stems from a one-tailed test.

(continued)

TABLE 8.1 (*continued*)

8.1.c. Enforcement-based specific deterrence

8.1.c.i. Qualitative measures of environmental behavior

Behavioral Measure	Category	Frequency (subsample conditional percentage)[a]		Test Statistic (p-value)[b]
		Below-Median-Deterrence Subsample	Above-Median-Deterrence Subsample	
Training: Quality	Lower: none, internal only, external only	NS[c]	NS[c]	0.052
	Higher: internal and external	NS[c]	NS[c]	(1.000)
TSS Treatment Presence	Absent	NS[c]	NS[c]	0.193
	Present	NS[c]	NS[c]	(1.000)
TSS Treatment Extent	Lesser: none, coagulation+sedimentation, sedimentation in effluent pond	NS[c]	NS[c]	0.266
	Greater: sedimentation in final clarifier, granular media filtration, membrane filtration	NS[c]	NS[c]	(1.000)
BOD Treatment Presence	Absent	NS[c]	NS[c]	0.029
	Present	NS[c]	NS[c]	(1.000)
BOD Treatment Extent	Lesser: none, solids removal, stabilization pond, trickling filter	NS[c]	NS[c]	0.522
	Greater: activated sludge, chemical oxidation, carbon adsorption	NS[c]	NS[c]	(0.949)
Treatment Upgrade: Presence	No	27	4	0.362
	Yes	58 (68.24)	5 (55.56)	(0.999)
Treatment Upgrade: Designed to Reduce Wastewater	No	40	4	0.107
	Yes	43 (51.81)	5 (55.56)	(1.000)

Monitoring Program	Lesser: none (except final discharge), treatment process, single-point sewer and treatment process	NS[c]		0.230 (1.000)
	Greater: multiple-point sewer and treatment process	NS[c]		
Audit: Team Composition	lesser: none, visiting corporate/consultants, teams of Facility employees	NS[c]		0.245 (1.000)
	Greater: teams of visiting corporate/ consultants and facility employees	NS[c]		
Audit: Classification Protocol	Lesser: none, noncompliance findings, noncompliance versus other, priority categories of vulnerability	5	65	0.587 (0.881)
	Greater: ranked serially by vulnerability for noncompliance	0 (0.00)	24 (26.97)	
External Consultants: Use	No	4	17	0.711 (0.692)
	Yes	5 (55.56)	70 (80.46)	
Requested Assistance from Wastewater Regulator	No	NS[c]		0.446 (0.989)
	Yes	NS[c]		
Compliance Assistance Meets Facility's Environmental Employees' Needs	Lesser: never, some of time, most of time	5	54	0.340 (1.000)
	Greater: always	4 (44.44)	26 (32.50)	
Importance Management Places on Role of General Plant Employees in Identifying and Correcting Conditions leading to Noncompliance	Lesser: somewhat important, important	0	16	0.383 (0.999)
	Greater: very important	5 (100.00)	75 (82.42)	
ISO 14001 Compliance	No	NS[c]		0.639 (0.809)
	Yes	NS[c]		

[a]The subsample conditional percentage is reported only for the category that represents either the presence of a behavioral measure or the greater extent of a behavioral measure.

[b]Test statistic applies to the comparison between the below-median-deterrence subsample and the above-median-deterrence subsample.

[c]NS indicates that at least one cell in the cross-tabulation includes three or fewer observations if the cell is populated.

(continued)

TABLE 8.1 (continued)

8.1.c.ii. Quantitative measures of environmental behavior

Behavioral Measure	Deterrence Subsample[a]	N	Mean	Sign of Difference in Means[b]	Test Statistic (p-value)[c]
Environmental Employees: Absolute	Below median	254	97.244	+	0.957
	Above median	26	139.192		(0.169)
Environmental Employees: Relative to Overall Employees	Below median	247	0.055	−	1.230
	Above median	26	0.047		(0.109)
Wastewater Employees: Absolute	Below median	90	12.372	−	−1.237
	Above median	5	4.600		(0.108)
Wastewater Employees: Relative to Overall Employees	Below median	87	0.107	−	−1.025
	Above median	5	0.026		(0.153)
Environmental Employees' Education	Below median	89	53.618	+	0.866
	Above median	5	73.200		(0.193)
Environmental Employees' Work Experience	Below median	90	13.611	−	−1.288
	Above median	5	11.000		(0.099)
Training Days	Below median	86	10.272	−	−1.279
	Above median	5	4.300		(0.101)
Training Attendance	Below median	89	90.371	−	−0.071
	Above median	5	82.800		(0.472)
Audit: Annual Count	Below median	240	5.163	+	0.255
	Above median	22	11.773		(0.399)
External Consultants: Annual Expenditures—Absolute	Below median	80	34,456	−	−0.908
	Above median	8	24,375		(0.182)
External Consultants: Annual Expenditures—Relative to Overall Employees	Below median	79	298.123	−	−1.413
	Above median	8	110.321		(0.079)
Management's Steps to Communicate Environmental Performance Goals and Targets to Facility Employees (count)	Below median	92	3.326	+	0.092
	Above median	5	3.400		(0.463)
Management's Steps to Communicate Environmental Performance Goals and Targets to External Parties (count)	Below median	92	2.370	−	−1.439
	Above median	5	1.400		(0.075)

[a]Deterrence categories: below median deterrence, above median deterrence.

[b]The difference in mean values = (mean of the above-median-deterrence subsample) − (mean of the below-median-deterrence subsample).

[c]Test statistic applies to the comparison between the below-median-limit subsample and above-median-limit subsample; it reports the Wilcoxon two-sample test statistic in

8.1.d. Enforcement-based general deterrence

8.1.d.i. Qualitative measures of environmental behavior

Behavioral Measure	Category	Frequency (subsample conditional percentage)[a]		Test Statistic (p-value)[b]
		Below-Median-Deterrence Subsample	Above-Median-Deterrence Subsample	
Training: Quality	Lower: none, internal only, external only	31	25	0.199
	Higher: internal and external	21 (40.38)	20 (44.44)	(1.000)
TSS Treatment Presence	Absent	7	4	0.225
	Present	44 (86.27)	40 (90.91)	(1.000)
TSS Treatment Extent	Lesser: none, coagulation + sedimentation, sedimentation in effluent pond	28	18	0.680
	Greater: sedimentation in final clarifier, granular media filtration, membrane filtration	23 (45.10)	26 (59.09)	(0.744)
BOD Treatment Presence	Absent	13	5	0.703
	Present	38 (74.51)	40 (88.89)	(0.706)
BOD Treatment Extent	Lesser: none, solids removal, stabilization pond, trickling filter	24	17	0.454
	Greater: activated sludge, chemical oxidation, carbon adsorption	27 (52.94)	28 (62.22)	(0.986)
Treatment Upgrade: Presence	No	12	19	0.790
	Yes	36 (75.00)	27 (58.70)	(0.560)

TABLE 8.1 (*continued*)

8.1.d. Enforcement-based general deterrence

8.1.d.i. Qualitative measures of environmental behavior

Behavioral Measure	Category	Frequency (subsample conditional percentage)[a]		Test Statistic (p-value)[b]
		Below-Median-Deterrence Subsample	Above-Median-Deterrence Subsample	
Treatment Upgrade: Designed to Reduce Wastewater	No	17	27	1.043
	Yes	29	19	(0.227)
		(63.04)	(41.30)	
Monitoring Program	Lesser: none (except final discharge), treatment process, single-point sewer and treatment process	28	20	0.511
	Greater: multiple-point sewer and treatment process	23	25	(0.956)
		(45.10)	(55.56)	
Audit: Team Composition	Lesser: none, visiting corporate/consultants, teams of facility employees	21	13	0.681
	Greater: teams of visiting corporate/consultants and facility employees	27	31	(0.743)
		(56.25)	(70.45)	
Audit: Classification Protocol	Lesser: none, noncompliance findings, noncompliance versus other, priority categories of vulnerability	37	33	0.203
	Greater: ranked serially by vulnerability for noncompliance	14	10	(1.000)
		(27.45)	(23.26)	
External Consultants: Use	No	7	14	0.804
	Yes	43	32	(0.537)
		(86.00)	(69.57)	

Requested Assistance from Wastewater Regulator	No	32	39	0.795
	Yes	16 (33.33)	8 (17.02)	(0.553)
Compliance Assistance Meets Facility's Environmental Employees' Needs	Lesser: never, some of time, most of time	30	29	0.176
	Greater: always	14 (31.82)	16 (35.56)	(1.000)
Importance Management Places on Role of General Plant Employees in Identifying and Correcting Conditions leading to Noncompliance	Lesser: somewhat important, important	10	6	0.307
	Greater: very important	41 (80.39)	39 (86.67)	(1.000)
ISO 14001 Compliance	No	41	38	0.131
	Yes	5 (10.87)	6 (13.64)	(1.000)

[a]The subsample conditional percentage is reported only for the category that represents either the presence of a behavioral measure or the greater extent of a behavioral measure.

[b]Test statistic applies to the comparison between the below-median–deterrence subsample and the above-median–deterrence subsample.

(continued)

TABLE 8.1 *(continued)*

8.1.d.ii. Quantitative measures of environmental behavior

Behavioral Measure	Deterrence Subsample[a]	N	Mean	Sign of Difference in Means[b]	Test Statistic (p-value)[c]
Environmental Employees: Absolute	Below median	163	101.779	–	–0.712
	Above median	117	100.248		(0.238)
Environmental Employees: Relative to Overall Employees	Below median	157	0.053	+	0.556
	Above median	116	0.057		(0.289)
Wastewater Employees: Absolute	Below median	52	13.933	–	–0.847
	Above median	43	9.581		(0.198)
Wastewater Employees: Relative to Overall Employees	Below median	49	0.104	–	–0.431
	Above median	43	0.101		(0.333)
Environmental Employees' Education	Below median	51	47.235	+	1.981
	Above median	43	63.465		(0.024)
Environmental Employees' Work Experience	Below median	50	13.140	+	0.535
	Above median	45	13.844		(0.297)
Training Days	Below median	47	7.689	+	1.062
	Above median	44	12.352		(0.144)
Training Attendance	Below median	50	89.720	+	0.159
	Above median	44	90.250		(0.437)
Audit: Annual Count	Below median	152	5.316	+	–1.516
	Above median	110	6.273		(0.065)
External Consultants: Annual Expenditures— Absolute	Below median	45	32,022	+	–0.324
	Above median	43	35,128		(0.373)
External Consultants: Annual Expenditures— Relative to Overall Employees	Below median	44	256.836	+	–0.154
	Above median	43	305.429		(0.439)
Management's Steps to Communicate Environmental Performance Goals and Targets to Facility Employees (count)	Below median	52	3.327	+	–0.196
	Above median	45	3.333		(0.422)
Management's Steps to Communicate Environmental Performance Goals and Targets to External Parties (count)	Below median	52	2.385	–	–0.468
	Above median	45	2.244		(0.320)

[a]Deterrence categories: below median deterrence, above median deterrence.

[b]The difference in mean values = (mean of the above-median-deterrence subsample) – (mean of the below-median-deterrence subsample).

[c]Test statistic applies to the comparison between the below-median-limit subsample and above-median-limit subsample; it reports the Wilcoxon two-sample test statistic in the form of a normal approximation; the associated p-value stems from a one-tailed test.

TABLE 8.1 *(continued)*

8.1.e. Injunctive relief-based specific deterrence

Behavioral Measure	Deterrence Subsample	N	Mean	Sign of Difference in Means[a]	Test Statistic (p-value)[b]
Environmental Employees: Absolute	Below median	276	NS[c]	–	−1.622 (0.053)
	Above median	1	NS[c]		
Environmental Employees: Relative to Overall Employees	Below median	269	NS[c]	–	−1.187 (0.118)
	Above median	1	NS[c]		
Audit: Annual Count	Below median	260	NS[c]	–	−1.600 (0.055)
	Above median	1	NS[c]		

8.1.f. Injunctive relief general deterrence

Behavioral Measure	Deterrence Subsample	N	Mean	Sign of Difference in Means[a]	Test Statistic (p-value)[b]
Environmental Employees: Absolute	Below median	213	97.653	+	−0.129 (0.449)
	Above median	64	98.109		
Environmental Employees: Relative to Overall Employees	Below median	206	0.059	–	−0.228 (0.410)
	Above median	64	0.036		
Audit: Annual Count	Below median	200	5.855	–	−1.400 (0.081)
	Above median	61	5.344		

8.1.g. SEP general deterrence

Behavioral Measure	Deterrence Subsample	N	Mean	Sign of Difference in Means[a]	Test Statistic (p-value)[b]
Environmental Employees: Absolute	Below median	249	97.446	+	0.179 (0.429)
	Above median	28	100.54		
Environmental Employees: Relative to Overall Employees	Below median	242	0.056	–	−0.187 (0.426)
	Above median	28	0.036		
Audit: Annual Count	Below median	235	5.804	–	−0.581 (0.281)
	Above median	26	5.115		

[a]The difference in mean values = (mean of the above-median-deterrence subsample) − (mean of the below-median-deterrence subsample).

[b]Test statistic applies to the comparison between the below-median-deterrence subsample and the above-median-deterrence subsample; it reports the Wilcoxon two-sample test as a normal approximation; p-value stems from a one-tailed test.

[c]NS indicates that at least one subsample includes three or fewer observations.

8.2.1.2.2. Hypothesis Testing: Enforcement-Related Deterrence. We next explore enforcement-related deterrence. The two-sample test results also indicate that the influence of enforcement-related specific deterrence varies across the forms of environmental behavior. Consistent with our hypothesis, enforcement-related specific deterrence positively influences the number of environmental employees—in absolute terms and relative to overall employees—and environmental employees' education. Inconsistent with our hypothesis, enforcement-related specific deterrence negatively influences other forms of environmental behavior: wastewater employees in absolute terms, environmental employees' work experience, training days, external consultant expenditures relative to overall employees, and management steps to communicate environmental performance goals and targets to external parties.[7] Providing no evidence for our hypothesis, enforcement-related specific deterrence does not appear to influence the remaining forms of environmental behavior. Table 8.1.c fully tabulates the test results.

All correlations indicate that enforcement-related specific deterrence does not appear to influence quantitative measures of environmental behavior (p > 0.10).

Similar to enforcement-related specific deterrence, the two-sample test results indicate that the influence of enforcement-related general deterrence varies across the forms of environmental behavior. Consistent with our hypothesis, enforcement-related general deterrence positively influences environmental employees' education. Inconsistent with our hypothesis, enforcement-related general deterrence negatively influences audit count.[8] Enforcement-related general deterrence does not appear to influence the remaining forms of environmental behavior. Table 8.1.d fully tabulates the test results.

All but one of the correlations indicate that enforcement-related general deterrence does not influence (significantly) the quantitative measures of environmental behavior. As the single exception, consistent with our hypothesis, enforcement-related general deterrence positively influences environmental employees' education ($\rho = 0.212$, $p = 0.040$).

8.2.1.2.3. Hypothesis Testing: Injunction-Related Deterrence. We next explore injunction-related deterrence. The two-sample test results shown in Table 8.1.e indicate that the influence of injunction-related specific deterrence negatively influences all forms of environmental behavior that is measured in each of three calendar years: environmental employees—in absolute terms and relative to overall employees—and audit count, inconsistent with our hypothesis. This influence is statistically significant or at least nearly so for all three behavioral forms.[9] However, these results are questionable since they are based on a comparison of subsamples in which the above-median-deterrence subsample includes a single observation (that is, one facility in a single year received an injunctive relief sanction). Consistent with this concern, the correlations indicate that injunction-related specific deterrence does not influence (significantly) any of these behavioral forms.

The two-sample test results shown in Table 8.1.f indicate that, similar to injunction-related specific deterrence, injunction-related general deterrence negatively influences the audit count, inconsistent with our hypothesis. However, the two-sample test results indicate that injunction-related general deterrence does not appear to influence the number of environmental employees—either in absolute terms or relative to overall employees. The correlations are consistent with this latter conclusion since they indicate that injunction-related general deterrence does not influence (significantly) the examined behavior.

8.2.1.2.4. Hypothesis Testing: SEP-Related Deterrence   Finally, we explore SEP-related deterrence. Since none of the survey respondents faced a SEP during the sample period, our analysis cannot examine the influence of SEP-related specific deterrence, so we focus instead on SEP-related general deterrence. The two-sample test results in Table 8.1.g indicate that SEP-related general deterrence does not appear to influence any of the examined forms of environmental behavior. The correlations support this conclusion since they indicate that SEP-related general deterrence does not influence (significantly) the examined behavior.

8.2.1.3. Evidence of Deterrence.   Based on this abundance of analytical results—two-sample tests and correlations, we assess the weight of evidence to support or reject our hypothesis that greater deterrence spurs better environmental behavior, that is, government interventions effectively prompt better environmental behavior. In general, we do not comment on results that neither support nor reject our hypothesis.

First, we assess the number of environmental employees. With the exception of enforcement-related general deterrence, greater deterrence spurs facilities to hire more environmental employees, especially relative to the number of overall employees. This evidence strongly supports our hypothesis. (In drawing this conclusion, we dismiss the questionable negative effect of injunction-related specific deterrence on the number of environmental employees.)

Second, we assess the number of employees devoted to wastewater management. Inconsistent with our hypothesis, greater enforcement-related specific deterrence spurs facilities to hire fewer wastewater workers (in absolute terms), but the effect is only marginally significant ($p = 0.108$).

Moreover, in terms relative to overall facility employees, the effect is insignificant.

Third, the effect of deterrence on employees' education depends on the type of government intervention. While greater inspection-related specific deterrence and general deterrence lower education, greater enforcement-related specific deterrence and general deterrence increase education. At least in the case of specific deterrence, perhaps the information provided by inspections is so useful that it reduces the need for facilities to employ better-educated environmental employees. In contrast, enforcement proceedings provide sufficiently limited information that the expected deterrent effect dominates.

Fourth, similar to environmental employees' education, greater inspection-related general deterrence appears to spur facilities to hire environmental employees with less work experience. Yet, in contrast to education, greater enforcement-related specific deterrence also appears to spur facilities to hire less-experienced workers. Both results provide evidence against our hypothesis. Perhaps the problems identified as being responsible for the government interventions concerned are of the sort that can be remedied through projects that do not require experienced workers, such as the basic construction of treatment facilities. If facilities are budget constrained, then an increase in treatment facilities may prompt a decrease in the budget resources available for experienced workers.

Fifth, deterrence negatively affects facilities' provision of training quantity regardless of the quantity dimension. Greater deterrence—in the form of enforcement-related specific deterrence—appears to spur facilities to provide fewer days of training. Similarly, greater deterrence—in the form of inspection-related general deterrence—appears to spur facilities to provide training to a lesser portion of their environmental employees. Both results are inconsistent with our hypothesis.

Sixth, greater inspection-related specific deterrence appears to spur facilities to perform more audits, yet greater inspection-related general deterrence and enforcement-related specific deterrence appears to spur facilities to perform fewer audits. Greater enforcement-related general deterrence and greater injunction-related general deterrence also appears to spur facilities to perform fewer audits. While the first result demonstrates that inspections performed at a particular facility spurs that particular facility to action, the second result indicates that the increased threat of an inspection based on the experiences of other facilities actually prompts a particular facility to reduce its effort. As noted in Section 8.1.2, in the case of general deterrence, perhaps an individual facility believes that agencies have used their limited monitoring resources to pursue inspections against other facilities, reducing the chance that the individual facility will be inspected. A similar explanation may apply to the third and fourth results, which indicate that facilities appear to use the experiences of other facilities to behave in ways apparently contrary to the notion of deterrence. We dismiss the questionable effect of injunction-related specific deterrence. None of these conclusions (except the last) are supported by the correlations since they indicate no link from deterrence to audit decisions. Thus, the evidence of contrary general deterrence effects is limited.

Seventh, the effect of deterrence on external consultant use depends on the form of deterrence. Greater inspection-related general deterrence spurs facilities to spend more on external consultants, especially in relative terms, in support of our hypothesis. Yet, greater inspection-related specific deterrence prompts facilities to spend less on external consultants in absolute terms, providing evidence against our hypothesis. Perhaps the information gleaned by the facility from recent inspections reduces the need to hire environmental consultants. As important, in relative terms, no link to external consultant expenditures proves statistically significant. Thus, in this case, it appears important to adjust for facility size. Similar to inspections, greater enforcement-related specific deterrence prompts facilities to spend less on external consultants even in relative terms, inconsistent with our hypothesis. While less likely, similar to inspections, perhaps recent enforcement encounters provide information that reduces the need to hire external consultants.

Eighth, the effect of deterrence on management efforts to communicate environmental performance goals and targets depends on the recipient of these communications. Greater deterrence—in the form of inspection-related general deterrence—spurs facility management to employ more steps to communicate environmental performance goals and targets to facility employees, in support of our hypothesis. However, greater deterrence—in the form of enforcement-related specific deterrence—spurs facility management to employ fewer steps to communicate environmental performance goals and targets to external parties, providing evidence against our hypothesis. Perhaps facilities perceive such external communications as increasing the risk that information that adversely reflects on behavior or compliance status may fall into the hands of government officials responsible for deciding whether to initiate future interventions.

For all the remaining forms of environmental behavior, the analytical results reveal no identifiable link from deterrence to behavioral decisions. Thus, when considering many dimensions of facility environmental management, greater deterrence neither spurs nor impedes better behavior.

Overall, these results provide limited evidence to support our hypothesis. Instead, most results indicate that interventions do not affect environmental behavior. Some results indicate that greater deterrence may appear to spur worse environmental behavior for reasons described earlier.

8.2.2. Multivariate Regression Analysis

To assess our hypothesis further, we employ our second empirical approach: multivariate regression analysis.

8.2.2.1. Description of Multivariate Approach. Multivariate regression analysis represents a commonly used statistical tool for establishing a functional relationship between an outcome, such as a facility's environmental behavioral decision, and a set of explanatory factors, such as the size of the facility. In contrast to theoretical analysis, multivariate regression analysis estimates the functional relationship based on relevant data. In the context of estimation, the outcome represents the dependent variable and the explanatory factors represent the independent variables; the former is so named because it depends on the independent variables. For our analysis here, we consider the most simple functional relationship—a linear relationship.

As one of its strengths, multivariate regression analysis is able to explore a functional relationship involving multiple factors or variables (so named "multivariate"). In contrast, two-sample testing and an assessment of pairwise correlations are able to explore a functional relationship involving only a single factor. In the appendix to Chapter 5, we explore whether environmental behavioral decisions vary based on certain factors: chemical manufacturing subsector, EPA region, facility size, and calendar year. In this chapter, we incorporate these multiple factors simultaneously into our multivariate analysis. (We are able to incorporate calendar year only in the case of environmental behavior that is measured for each calendar year between 1999 and 2001.) More important, we incorporate the multiple dimensions of deterrence simultaneously into our multivariate analysis: federal inspections, state inspections, informal enforcement actions, administrative formal nonsanction actions, civil formal nonsanction actions, administrative fines, and civil fines. When exploring forms of environmental behavior measured for each calendar year between 1999 and 2001, we also incorporate administrative injunctions, civil injunctions, administrative SEPs, and civil SEPs.

By incorporating multiple factors into our multivariate regression analysis, we are able to isolate the effect of each deterrence type, independent of the influence of other factors. Clearly, we wish to identify the independent effect of each deterrence type. However, we are not as interested in identifying the independent effects of subsector, EPA region, facility size, and calendar year. Given our focus on deterrence, these other variables represent control factors because we control for their influences in order to isolate the effects of our key variables—deterrence types.

In Chapter 5, we explore the relationship between environmental behavior and effluent limits. However, we are reluctant to incorporate this particular factor, despite its importance, into our multivariate regression analysis because several facilities face neither total suspended solids (TSS) nor biological oxygen demand (BOD) effluent limits. By incorporating this factor, we would need to further reduce our samples, which are already small, for the purposes of multivariate analysis involving numerous deterrence factors. Rather than exploring the effect of effluent limits on behavior further, we rest on our analysis provided in Chapter 5. In contrast, the sample sizes used to estimate environmental performance in Chapter 9 are much larger. Moreover, in Chapter 9, we are able to connect pollutant-specific, pollution-measurement-specific limits (for example, TSS concentration limits) to pollutant-specific, measurement-specific discharges. Both fortunate circumstances induce us to incorporate gladly discharge limits into our multivariate regression analysis within Chapter 9, which provides additional information about the effect of discharge limits on environmental performance.

As noted many times, the data on environmental behavior divide into two categories: quantitative measures, such as external consulting expenditures, and qualitative measures, such as the presence of ISO 14001 compliance. However, in the context of multivariate analysis, this division is not fully accurate for certain forms of behavior. In particular, some forms are better categorized as count data: annual count of audits, count of management's steps to communicate environmental performance goals and targets to facility employees, and count of management's steps to communicate environmental performance goals and targets to external parties. We transform all qualitative measures into dichotomous measures, as described in Chapter 5.

Given this clarification, we explore three types of behavioral data: quantitative, dichotomous, and count. When exploring quantitative behavioral forms, we employ the most common type of multivariate regression analysis: ordinary least squares regression. While other types of multivariate analysis may be more appropriate in certain cases, ordinary least squares is sufficient for our purposes. Of these quantitative behavioral forms, we measure two forms—absolute number of environmental employees and ratio of environmental employees to overall employees—in each of three calendar years between 1999 and 2001. Thus, we possess longitudinal or panel data on these forms. In order to exploit and accommodate this panel data structure, we employ a fixed-effects estimator, which incorporates a new set of control factors: an indicator factor for each facility. This factor controls for heterogeneity (that is, characteristic differences) across the facilities that is fixed over time for each facility. This approach seems appropriate when each facility is different in some way that is not captured by the other control factors (for example, facility size) and this difference does not change over time (for example, the manager of each specific facility is different in his or her own way and this manager runs each facility over the entire three-year period).[10] When exploring qualitative behavioral forms, we employ the probit estimator, which represents the most commonly used method for exploring qualitative dependent variables that represent two categories.[11] When exploring count behavioral forms, we employ the Poisson estimator, which represents the most commonly used method for exploring count-dependent variables.[12] Moreover, we possess panel data on audit counts. In order to exploit and accommodate this panel data structure, we employ a random-effects Poisson estimator, which allows for heterogeneity across the facilities that relates to the unknown aspects of the relationship between audit counts and the explanatory factors. This approach seems appropriate when each facility is different in some way that is not known to the researcher.[13]

Lastly, we describe our ability to incorporate the many dimensions of deterrence into our multivariate analysis based on the actual use of the various government interventions: inspections, informal enforcement actions, and so on. Clearly, we are not able to incorporate a deterrence factor whenever the underlying government intervention is not performed in our sample, or during our sample period in the case of general deterrence. When examining calendar-year behavior, we are not able to incorporate civil-fine-related specific deterrence and civil-fine-related general deterrence. When examining the influence of injunctions and SEPs on these same behavioral forms, we are not able to incorporate these deterrence factors: civil injunction–related specific deterrence, administrative SEP–related specific deterrence, civil SEP–related specific deterrence, civil injunction–related general deterrence, administrative SEP–related general deterrence, and civil formal nonsanction–related specific deterrence. When estimating "preceding twelve-month period" behavior, we not able to incorporate administrative fine–related specific deterrence, civil fine–related specific deterrence, and civil fine–related general deterrence. When examining particular "preceding twelve-month period" behavioral forms that represent qualitative outcomes (for example, presence versus absence of TSS treatment), we are not able to incorporate additional deterrence factors, as noted in the following sections for each relevant case. When estimating "preceding three-year period" behavior, we are not able to incorporate these deterrence factors: civil formal nonsanction specific deterrence, administrative fine–related specific deterrence, civil fine–related specific deterrence, and civil fine–related general deterrence.

8.2.2.2. Reporting and Interpreting Multivariate Regression Results. Our use of multivariate regression analysis generates many sets of empirical results, including estimates of coefficient magnitudes and the associated standard errors. The coefficient magnitudes represent the multiplicative factors applied to the independent variables within the estimated linear functional relationship between the dependent variable and the independent variables. (For example, m represents the coefficient magnitude in the following functional relationship between a dependent variable y and a single independent variable x: $y = b + mx$.) The standard error associated with a coefficient reflects the "noisiness" of the coefficient magnitude and measures the "tightness" of the coefficient estimate. By comparing the coefficient magnitude to its standard error, the multivariate regression analysis is able to assess the statistical significance of each independent variable's coefficient estimate, as reflected in each coefficient's p-value. This assessment stems from a null hypothesis of a zero coefficient magnitude, which implies that a particular independent variable has no influence on the dependent variable. We deem a coefficient estimate statistically significant when our analysis is able to reject the null hypothesis of a zero coefficient at a significance level of 10 percent or less, as revealed by a p-value of 0.10 or less. We deem a coefficient estimate statistically insignificant when our analysis is not able to reject the null hypothesis of a zero coefficient at a significance level of 10 percent or less, as revealed by a p-value greater than 0.10.

We report and interpret only the coefficients relating to the influence of deterrence on environmental behavior. We use common notation to identify the coefficient magnitude (β) and the level of statistical significance (p).

We explore a variety of environmental behavioral measures. First, we explore the quantity and quality of labor effort devoted to environmental management. We measure quantity based on two classifications: overall number of environmental employees and the number of employees devoted to wastewater management, in absolute terms and relative to the overall number of facility employees. We assess the effect of deterrence without and with data on injunctions and SEPs. We assess the effects of injunctions and SEPs only once these data are included. Multivariate results indicate that civil formal nonsanction specific deterrence negatively influences the absolute number of environmental employees, implying that greater deterrence spurs facilities to hire fewer environmental workers ($\beta = -14.5$, p=0.000). Otherwise, deterrence does not influence the absolute number of environmental employees. The multivariate results also indicate that state inspection–related general deterrence and civil formal nonsanction–related general deterrence positively influence the relative number of environmental employees, while administrative fine–related specific deterrence and general deterrence both negatively influence the relative number of environmental employees (β=0.008, p=0.024; β=0.631, p=0.000; β= −9.89E–7, p=0.001; β=−1.14E–5, p=0.010; respectively). Collectively, these results indicate that greater deterrence from interventions less severe than fines spurs facilities to hire more environmental employees (relative to overall employees), yet greater deterrence from fines actually prompts facilities to employ fewer environmental workers (relative to overall employees). When including data on injunction and SEPs, the multivariate results indicate that neither injunctions nor SEPs influence the number of environmental employees in absolute or relative terms.

In addition, we assess the influence of deterrence on the number of wastewater employees. Multivariate results indicate that fine-based general deterrence negatively influences the absolute number of wastewater employees (β=−1.026, p=0.060), implying that greater deterrence spurs facilities to hire fewer wastewater employees, contrary to our hypothesis. Perhaps the problems that prompted the imposition of fines can be remedied through projects that do not require wastewater employees. If facilities are budget constrained, then expenditures on these new projects may prompt a decrease in the budget resources available for wastewater employees. Otherwise, deterrence does not affect the number of wastewater employees—in absolute or relative terms.

Second, we assess the influence of deterrence on the quality of labor effort based on employees' education and work experience. Multivariate results indicate that administrative formal nonsanction–based specific deterrence negatively influences environmental employee education (β= −14.491, p=0.050), implying that greater deterrence spurs facilities to hire fewer college-educated environmental employees, contrary to our hypothesis. Perhaps the problems that prompted the imposition of formal nonsanctions can be remedied through projects that do not require college-educated environmental employees. If facilities are budget constrained, then expenditures on these new projects may prompt a decrease in the budget resources available for college-educated employees. Otherwise, deterrence does not influence education. Multivariate results also reveal that both administrative formal nonsanction–based specific deterrence and informal enforcement–based general deterrence positively influence work experience (β= 2.195, p=0.061; β=45.883, p=0.008, respectively), yet federal inspection–related general deterrence negatively influences work experience (β=−28.948, p=0.081). Thus, greater deterrence from enforcement spurs facilities to employ more-experienced workers, yet greater deterrence from inspections prompts facilities to employ less-experienced workers. Perhaps the information provided by inspectors reduces the need to employ experienced workers. Otherwise, deterrence is not influential.

We also assess the effect of deterrence on other measures of labor effort quality, which are based on the provision of training, its quality, and its quantity. Multivariate results indicate that administrative formal nonsanction–based general deterrence negatively influences training quality ($\beta = -190.195$, $p = 0.000$), implying that greater deterrence spurs facilities to offer lower-quality training. As noted in Section 8.1.2, perhaps an individual facility believes that agencies have used their limited enforcement resources to pursue formal actions against other facilities, reducing the chance that the individual facility will be targeted for enforcement. Otherwise, deterrence does not influence training quality.[14] Multivariate results also reveal that deterrence does not influence the quantity of training as measured by either days of training or degree of attendance.

Third, we assess the effect of deterrence on the provision of capital equipment, as measured by the use of treatment technologies and the extent of this treatment. Multivariate results indicate that deterrence influences neither the presence nor the extent of TSS treatment.[15] Multivariate results also indicate that deterrence does not influence the extent of BOD treatment extent. However, administrative formal nonsanction–based general deterrence positively influences BOD treatment presence ($\beta = 202.660$, $p = 0.000$), implying that greater deterrence spurs facilities to employ BOD treatment.[16]

We also assess the influence of deterrence on upgrades to treatment technologies. Multivariate results indicate that federal inspection–based specific deterrence and administrative formal nonsanction–based specific deterrence negatively influence treatment upgrades ($\beta = -1.389$, $p = 0.077$; $\beta = -0.413$, $p = 0.073$, respectively), implying that greater deterrence impedes upgrades. These negative links disappear once the analysis focuses on upgrades designed to reduce wastewater discharges. Thus, it appears important to account for the intention of an upgrade. Otherwise, deterrence is not influential.[17]

Fourth, we assess the influence of deterrence on facilities' efforts to understand their environmental management concerns. Consider the influence of deterrence on the extent of a facility's monitoring program. Multivariate results indicate that informal enforcement–based general deterrence positively influences yet state inspection–based specific deterrence negatively influences the extent of monitoring ($\beta = 8.269$, $p = 0.066$; $\beta = -0.332$, $p = 0.019$, respectively), implying that greater deterrence from enforcement spurs facilities to increase the extent of their monitoring yet greater deterrence from inspections prompts facilities to decrease their monitoring extent. Perhaps inspections provide information that is so useful that facilities feel a weaker need for monitoring. Otherwise, deterrence is not influential.[18]

More important, we assess the influence of deterrence on facilities' effort to audit their operations. We assess both the quantity and quality of audits, starting with quantity. Multivariate results indicate that civil formal nonsanction–based specific deterrence positively influences the annual count of audits ($\beta = 0.156$, p=0.000). Multivariate results also indicate that both administrative injunction–related general deterrence and civil SEP–related general deterrence positively influence the count of audits ($\beta = 473$, p=0.026; $\beta = 422$, p=0.030, respectively). Thus, greater deterrence spurs facilities to conduct more audits per year, in support of our hypothesis.[19] Other forms of deterrence do not influence audit count.

We also assess the influence of deterrence on the quality of audits based on the audit team's composition and the extent of the audit classification protocol. Multivariate results reveal that no link exists from deterrence to audit team composition. However, the results demonstrate that federal inspection–related general deterrence positively influences yet fine-related general deterrence negatively influences the extent of the audit classification protocol ($\beta = 10.557$, p=0.031; $\beta = -0.112$, p=0.016, respectively). Thus, greater deterrence from inspections spurs facilities to implement a more extensive audit classification protocol, while greater deterrence from the most severe form of enforcement induces facilities to implement a less extensive audit classification protocol. Otherwise, deterrence appears ineffective.[20]

Fifth, we assess the influence of deterrence on facilities' efforts to seek insight from others regarding their environmental management. Consider the effect of deterrence on the use of external consultants. Multivariate results indicate that deterrence does not affect the decision whether or not to hire an external consultant. However, multivariate results indicate that deterrence affects the annual amount of money spent on external consultants in absolute terms but not relative to a facility's size. The results reveal that administrative formal nonsanction–based general deterrence positively influences absolute expenditures on external consultants ($\beta = 3,694,976$, p= 0.080). Yet the results also demonstrate that general deterrence stemming from both informal enforcement and civil formal nonsanctions, along with specific deterrence from state inspections negatively influences external consultant expenditures ($\beta = -1,255,835$, p=0.065; $\beta = -6,355,458$, p=0.038; $\beta = -11,492$, p=0.063, respectively). Thus, greater deterrence both spurs and impedes facilities' spending on external consultants. Other forms of deterrence do not influence absolute consultant expenditures.

We assess the influence of deterrence on facilities' efforts to gain insight from their regulators, in contrast to hired consultants. Multivariate results indicate that fine-based general deterrence positively influences these efforts, implying that greater deterrence stemming from the most severe form of enforcement spurs facilities to seek insight from their regulators, which should reassure the EPA and state regulators. Otherwise, deterrence is not influential.

We also assess the influence of deterrence on the success of facilities' efforts to gather information and insight as measured by how frequently compliance assistance materials meet the day-to-day needs of facilities' environmental employees. Multivariate results indicate that deterrence does not affect the frequency of this success unless the analysis assesses the effect of sanctions based on the count of sanctions rather than the monetary value of sanctions. Given this modification, informal enforcement–based general deterrence positively influences yet administrative formal nonsanction–based general deterrence negatively influences the success of compliance assistance materials ($\beta = 39.9$, p = 0.083; $\beta = -73.0$, p = 0.084). Thus, greater deterrence from informal enforcement spurs facilities to secure more useful compliance assistance materials, yet greater deterrence from formal enforcement induces facilities to secure less useful compliance assistance materials.

Sixth, we assess the influence of deterrence on facilities' degree of concern over environmental management and to what extent these concerns translate into goals that are communicated to the facility's employees. Consider the effect of deterrence on the level of importance management places on the role of general plant employees in identifying and correcting conditions that may lead to noncompliance. Multivariate results indicate that administrative formal nonsanction–based general deterrence positively influences the level of managerial importance, implying that greater deterrence prompts greater concern over environmental management ($\beta = 177$, p = 0.000), consistent with our hypothesis. Otherwise, deterrence is not influential.[21]

We also assess the influence of deterrence on the degree of effort expended by management to communicate environmental performance goals and targets to facility employees. Multivariate results indicate that deterrence does not influence the degree of managerial effort.

Seventh, we assess the influence of deterrence on management's communication of environmental performance goals and targets to external parties, in contrast to the communication to facility employees. Multivariate results indicate that informal enforcement–based general deterrence negatively influences the degree of managerial effort ($\beta = -4.13$, p = 0.043), implying that greater deterrence induces facility management to take fewer steps to communicate to external parties, contrary to our hypothesis. Perhaps an increase in the likelihood of informal enforcement prompts facilities to fear more greatly that external communications may fall into the hands of regulatory officials, increasing the likelihood of future government interventions. Otherwise, deterrence is not influential.

Eighth, we assess the effect of deterrence on ISO 14001 compliance. Multivariate results indicate that informal enforcement–based specific deterrence positively influences compliance ($\beta = 0.453$, $p = 0.097$), implying that greater deterrence spurs facilities to comply with ISO 14001 standards, in support of our hypothesis. Other forms of deterrence do not affect ISO 14001 compliance.[22]

Overall, these multivariate results provide only limited, if not mixed, evidence to support our hypothesis that government interventions prompt better environmental behavior.

8.3. SUMMARY

In this chapter, we evaluate the link between government interventions and facility behavior. To set the stage for our analysis, we distinguish between specific and general deterrence. We use two empirical approaches to test our hypothesis that greater deterrence spurred by government interventions prompts better environmental behavior. The first is to split the sample of facilities into two subsamples based on the degree of deterrence faced by an individual facility and then explore whether behavior is better when deterrence is greater. The second uses multivariate regression analysis to estimate a functional relationship between a specific behavioral decision and explanatory factors, focusing on factors relating to government interventions.

The influence of government interventions on behavior differs depending on whether we test for specific or general deterrence and on the form of behavior involved. We conclude, among other things, that the influence of both inspection-related specific deterrence and inspection-related general deterrence varies across the many forms of environmental behavior we identify. The same is true for the influence of enforcement-related specific deterrence. Overall, the split-sample approach indicates that interventions do not affect environmental behavior, and some results indicate that greater deterrence may appear to worsen environmental behavior. Our multivariate regression analysis provides limited, if not mixed, evidence to support our hypothesis that government interventions prompt better environmental behavior.

CHAPTER NINE

Effect of Government Interventions on Environmental Performance

This chapter analyzes the effect of government interventions on environmental performance. In particular, the chapter attempts to answer the following policy-relevant research question: do government interventions help to induce better environmental performance? To answer this question, we attempt to link the use of government interventions to facilities' environmental performance in order to generate evidence that helps to answer our research question. We employ two empirical approaches. One approach surveys facilities regarding their perceptions of the effectiveness of government interventions at prompting better compliance with National Pollutant Discharge Elimination System (NPDES) discharge limits. The second approach employs multivariate regression analysis to estimate a functional relationship between actual discharges and government interventions, while distinguishing between specific deterrence and general deterrence.

9.1. PERCEPTIONS OF CHEMICAL MANUFACTURING FACILITIES ABOUT THE IMPACT OF GOVERNMENT INTERVENTIONS ON ENVIRONMENTAL PERFORMANCE

As the first empirical approach, we assess responses from our survey of chemical manufacturing facilities that provide insights into how chemical plant personnel perceive the effectiveness of different forms of government interventions, including inspections, monetary fines, injunctive relief, and supplemental environmental projects (SEPs).[1] With this assessment in mind,

we designed our survey to elicit the views of chemical plant personnel involved in compliance with the facility's Clean Water Act responsibilities on
the impact of government interventions on environmental performance.[2]

9.1.1. Inspections

As the first type of intervention, the survey inquires whether the respondents thought that government inspections are effective ways for inducing individual chemical facilities to comply with permitted water discharge
limits. Of those surveyed, 52.1 percent said "definitely yes." Another 35.6
percent said "probably yes." Only 10.9 percent said "definitely not" or
"probably not." Thus, the overwhelming majority of the respondents
perceived inspections to be an effective way to improve environmental
performance.[3]

As a follow-up to this question, the survey asks whether it matters that
inspections are performed by a state regulator or EPA. Of those surveyed,
21.5 percent said "definitely yes" and another 15.5 percent said "probably yes." In contrast, 33.6 percent said "definitely not," and another 27.9
percent said "probably not." As a further follow-up, the survey asked those
who said that it matters whether inspections are performed by state or federal regulators, which type of inspection is more effective. Of those who
said the source of the inspection matters, 59.6 percent said that state inspections are more effective. Only 18.1 percent said federal inspections are
more effective. Oddly, 21.6 percent said they did not know which was
more effective.

These results seem to indicate that the survey participants overwhelm-
ingly thought that inspections are effective deterrents. A majority of the
respondents thought it does not matter whether inspections are performed by state or federal officials. This is the expected response, assuming that federal and state agencies are equally likely to perform inspections and equally likely to follow up on the results of those inspections
with enforcement action. But of those who perceived a difference, more
believed that state inspections are more effective than those who believed
that federal inspections are more effective. These results are consistent
with the proposition that the proximity of the inspectors affects behavior: state inspectors are closer than federal inspectors. Alternatively, these
results support the proposition that regulated entities are loath to undermine their cooperative relationships with state regulators and that they
make greater efforts to improve their performance following a state inspection. Another explanation is that state agencies implement more vigorous
types of inspections. In contrast, these results are inconsistent with the
proposition that state regulators are more averse to following up inspec-

tions with rigorous enforcement for fear that a state with a reputation for strong environmental enforcement may become comparatively less attractive to new business than a state with a reputation for weak environmental enforcement.

9.1.2. Monetary Fines

To gauge the perceptions of the chemical industry respondents about the effectiveness of monetary fines on environmental performance, the survey asks a series of questions about different kinds of fines. The survey asks a general question about the effectiveness of fines in inducing compliance, and follows up that question by asking several other questions relating to the difference between administrative and civil fines and the difference between federal and state fines.

To assess whether regulated chemical industry facilities regard monetary sanctions as effective in improving environmental performance, the survey asks generally whether the imposition of monetary fines is an effective way for inducing individual chemical facilities to comply with permitted discharge limits. More than three-quarters of the respondents (77.4 percent) said that monetary fines are an effective deterrent. As important, more than 21 percent of the survey participants said that fines are definitely not or probably not effective at inducing individual chemical facilities to comply with their Clean Water Act responsibilities. The fact that a substantial percentage of the survey respondents did not view monetary fines as effective deterrents may be the result of a perception that monetary penalties tend to be small. In particular, perhaps fines tend to be low enough to make it cheaper for regulated facilities in some instances to violate applicable discharge limitations and run the risk of federal or state enforcement that culminates in the imposition of fines than to spend money reducing discharge levels to conform to NPDES permit limits. The 21 percent figure is almost twice the percentage of the respondents in our survey that found inspections not to be effective.

To determine whether the respondents view certain kinds of fines as likely to be more effective at inducing compliance than others, the survey asks whether it matters whether the monetary fines are levied by the federal or state governments, assuming the fines are the same size. Of the respondents, 53.8 percent said it definitely does not matter, and another 29.5 percent said it probably does not. Only 13.7 percent said that the identity of the government matters, a much smaller number than the 37 percent who said that the identity of the governmental entity performing inspections matters. The fact that only 13.7 percent of the respondents felt that the identity of the governmental entity levying the fine matters is consistent with what economic analysis would predict. If the size of the fine is the same, it should not matter who the enforcer is because a dollar owed to the federal government has the same impact on the regulated entity required to pay it as a dollar owed to the state government. In other words, given the explicit equality of the size of the fine assessed by each level of government, the noted result conforms to expectations that two fines of equal size should have the same impact on a facility's bottom line, regardless of who imposes the fine.

The survey also asks those who thought the identity of the government imposing the fine matters whether a fine imposed by state or federal governments is more effective. Of those answering this question, 25.4 percent said they thought state fines were more effective, while 21.6 percent said they thought federal fines were more effective. Oddly, 48.5 percent said they did not know.

The survey also inquires of those who said that it matters whether the fine is federal or state why they thought it matters. The answers lend some support to the conclusion that regulated entities perceive federal fines to have a greater adverse impact on the violator's business, even if the size of the fines is the same. One common thread in many of the answers is that federal fines tend to generate more adverse publicity than state fines and therefore have adverse effects that go beyond the direct impact of the fine itself on the violator's bottom line. Those who thought that state fines are more effective seem to have focused on the proximity of the regulators to the firm and the possibility that state fines may actually generate more adverse publicity locally, even if publicity at the national level is not as great.

In sum, the survey respondents seemed to view monetary fines as an effective deterrent, but as one that is comparatively less effective than inspections as a deterrent to noncompliance with NPDES discharge limits. They also viewed the identity of the government entity imposing the fine as less important than the identity of the entity conducting an inspection. Perhaps that is because, as regards specific deterrence, the respondents perceived state inspectors as more likely than federal inspectors to follow up with enforcement action, whereas once a monetary penalty has been imposed, the firm suffers (at least directly) to the same extent whether the fine is state or federal in origin. Of course, the adverse consequences of monetary fines for the sanctioned facility may very well extend beyond the obligation to pay the amount assessed to the government. Additional indirect effects might include a reduction in future sales if consumers react adversely to publicity about the fine by reducing purchases of the violator's products or services. This kind of collateral damage might be significant for some facilities.

As another follow-up question to the initial inquiry regarding the effectiveness of fines, the survey asked whether it matters whether a federal fine is administratively imposed or judicially imposed (that is, in the form of a civil fine), explicitly stating that the fines are both the same size. Of the respondents, 41.5 percent said definitely yes or probably yes. Of the respondents, 49.4 percent said definitely not or probably not. The former figure is a much higher number than the percentage who said that the identity of the government agency matters. Of those who said that the source of the federal fine matters, in response to the next survey question, 25.7 percent said that administrative fines are more effective, while 39.1 percent said that civil fines are more effective. Thus, there is a clear lack of consensus on whether the impact of an administrative fine differs from that of a civil fine. However, if a difference exists, the responses provide some support for civil fines dominating administrative fines. In general, the survey responses support the conclusion that the difference between federal administrative and federal civil monetary fines as deterrents seems more important to the respondents than the difference between federal and state fines.

The survey poses a follow-up question to those respondents who said that it matters whether a fine is imposed administratively or judicially. This follow-up question asks the respondents why it matters. Based on the provided reasons, it seems that one of the factors that made some respondents perceive judicially imposed fines to be comparatively stronger deterrents than administrative fines was fear of adverse publicity. That may be because the imposition of a fine by a federal court for Clean Water Act violations is more likely to wind up being reported in print media such as local newspapers than the imposition of a fine by the EPA.[4] Other factors also may be at work. Some respondents, for example, thought that company employees were at greater risk of individual liability, including criminal liability in a related or follow-up enforcement proceeding, from judicially imposed fines than from administratively imposed fines.

The respondents who reported that administratively imposed fines are more effective than judicially imposed fines provided reasons relating to the branding by the EPA. Perhaps these respondents reasoned that regulated entities have an ongoing relationship with federal and state regulatory agencies, but not with courts.

9.1.3. Injunctive Relief

To gauge the perceptions of the survey respondents about the effectiveness of injunctive relief at inducing environmental compliance, the survey inquires whether imposition of a federal injunctive relief sanction

is an effective way to induce individual chemical facilities to comply with
permitted water discharge limits. Of the respondents, 22.0 percent said
definitely yes, while 29.2 percent said probably yes. In contrast, 18.6
percent of the respondents said definitely not or probably not; 28.8 percent said they did not know.

The percentage of respondents who view injunctions as effective (51.4
percent)—those answering probably yes or definitely yes—is much lower
than the percentage of those who view monetary fines as effective (77.4
percent). To explore this point, we directly compare the effectiveness of
injunctions to the effectiveness of fines as reported by respondents. For
this comparison, we examine the two relevant responses—reported effectiveness of injunctions and reported effectiveness of fines—for each
respondent. Based on these two responses, we classify each respondent
as granting a higher level of effectiveness to fines, a higher level to injunctions, or an identical level to both. Our analysis reveals that the
proportion of the sample believing that fines are a more effective means
of inducing compliance with Clean Water Act discharge limits than injunctions (31.0 percent) is significantly greater than the proportion believing that injunctions are more effective (19.6 percent).[5] This difference
may be due to the perception of the respondents that injunctive remedies
do relatively little to penalize violators for past illegal conduct. Injunctions by definition are prospective in orientation, as compared to the
backward-looking aspect of monetary fines, which penalize facilities for
past violations.

9.1.4. Supplemental Environmental Projects

To gauge the perceptions of chemical industry respondents concerning
the effectiveness of SEPs at inducing environmental compliance, the survey inquires whether a federal requirement to complete a SEP is an effective way to induce individual chemical facilities to comply with permitted discharge limits. A total of 69.2 percent of the respondents said that
SEPs definitely are or probably are effective; 9.1 percent said that these
projects are definitely not an effective way to induce compliance, and
another 14.8 percent said they are probably not effective. Of the respondents, 5.7 percent did not know whether SEPs are an effective means of
inducing compliance.

A majority of the respondents view SEPs as effective deterrents. This
result would appear to conflict with occasional criticisms of this kind of
intervention as a means of improving environmental performance. However, the percentage of respondents viewing SEPs as effective—those an-

swering definitely yes or probably yes—is lower than for any of the other kinds of government interventions covered by the survey except for injunctions. Thus, even if the respondents view SEPs as an effective deterrent, they may perceive SEPs as a comparatively weak one. To explore this point, similar to the analysis we use for injunctions, we compare the effectiveness of SEPs to the effectiveness of fines as reported by respondents. For this comparison, we examine the two relevant responses for each respondent. Based on these two responses, we classify each respondent as granting a higher level of effectiveness to fines, a higher level to SEPs, or an identical level to both. Our analysis reveals that the proportion of our sample believing that fines are a more effective means of inducing compliance with Clean Water Act permit responsibilities than SEPs (33.2 percent) is not significantly greater than the proportion believing that SEPs are more effective than fines (27.1 percent).[6]

One interesting question remains: why do survey respondents perceive injunctions as even weaker deterrents than SEPs? One possibility is that SEPs typically are the product of negotiation between the enforcing authority and the regulated entity. The latter therefore is likely to have more input into the contents of a SEP than into the obligations imposed by an injunction. To the extent that some injunctions, like SEPs, are the product of negotiations between enforcing authorities and dischargers, this explanation for why respondents perceived injunctions as weaker deterrents than SEPs loses force. Perhaps the respondents feel that injunctions are more adversarial in nature than SEPs. Therefore, injunctions are more likely to generate hostility toward, or at least undermine cooperation with, regulators. Consequently, future violations are more likely to occur following the issuance of an injunction than following the issuance of a SEP. Conversely, if the formulation of a SEP signals the beginning of a more cooperative relationship between regulator and regulated entity, the latter may work harder to avoid future violations so as not to lose the goodwill that may have been generated by negotiation and issuance of the SEP.

9.2. MULTIVARIATE REGRESSION ANALYSIS

9.2.1. Analytical Framework

The second empirical approach employs multivariate regression analysis. In particular, it uses regression analysis to estimate a functional relationship between a specific performance outcome (for example, total suspended solids [TSS] absolute quantity-based discharges) and a set of explanatory factors, with a focus on the factors relating to government interventions.

We do not employ an empirical approach that splits the sample of facilities into two subsamples based on the degree of deterrence faced by an individual facility, as identified by the median degree of deterrence, and then explores whether performance is better when deterrence is greater by employing two-sample tests. As noted in Chapter 8, the multivariate regression analysis dominates this alternative empirical approach because the former is able to explore a functional relationship involving multiple factors, while two-sample testing is able to explore a functional relationship involving only a single factor. While both empirical approaches are useful, we prefer to focus on the stronger approach in order to provide a more varied use of the multivariate regression analysis by assessing multiple sample periods.

As our general hypothesis, we test whether greater deterrence prompts better environmental behavior. We interpret our test results within the following light: if performance improves as deterrence increases, then government interventions would appear effective at prompting better performance. In order to test this general hypothesis, we examine the effects of government intervention–based deterrence on the various measures of environmental performance as revealed by the estimated relationship between environmental performance and government intervention factors. We explore eleven types of government interventions: federal inspections, state inspections, informal enforcement actions, administrative formal nonsanctions, civil formal nonsanctions, administrative fines, civil fines, administrative injunctive relief, civil injunctive relief, administrative SEPs, and civil SEPs. We explore two forms of deterrence: specific deterrence and general deterrence. Section 8.1 in Chapter 8 describes the construction of these deterrence measures. In total, we explore the effects of twenty-two different deterrence measures on environmental performance outcomes.

Chapter 6 describes multiple forms of environmental performance. Consistent with this depiction, we explore two types of wastewater pollutant: total suspended solids (TSS) and biological oxygen demand (BOD); two forms of measurement: quantity based and concentration based; and two frames: absolute and relative to imposed discharge limits. In total, we explore eight forms of environmental performance. When exploring quantity-based absolute discharges, we adjust both discharges and limits with respect to facility-specific production levels, consistent with Chapter 6. As long as quantity-based limits reflect variation in production across facilities, by including the quantity-based discharge-limit level as an explanatory factor in our multivariate analysis, we may not need to adjust either discharges or limit levels. We explore this point by also estimating a functional relationship for quantity-based discharges without any production adjustment. In the process, we assess the robustness of our results. Consistent with this point, when estimating the functional relationship involving quantity-based relative discharges, we may or may not need to adjust discharge-limit levels with respect to production. Regardless, we clearly should not adjust the level of relative discharges (that is, the ratio of absolute discharges and discharge limit) since it represents the extent of compliance. Again, this additional estimation assesses the robustness of our results. (In all cases, the two sets of estimates are nearly identical, so we report details only for a single set of estimates: arbitrarily, those generated by adjusting for production.)

The multivariate regression analysis estimates the functional relationship between environmental performance outcomes and multiple explanatory factors. As the most important factors, we incorporate the multiple dimensions of deterrence simultaneously into our multivariate analysis. When incorporating administrative injunctions, civil injunctions, administrative SEPs, and civil SEPs, we must limit the sample period to those months when data on injunctions and SEPs are available, as described later in this chapter. As a complement to our empirical analysis in Chapter 6, we also incorporate the discharge-limit level and multiple dimensions of environmental behavior. The incorporation of behavioral measures depends on the sample period explored, as described below. By incorporating these key factors, we are able to explore more fully the effect of discharge limits on environmental performance (do tighter discharge limits lower absolute discharges, while raising discharge ratios?) and the effect of environmental behavior on environmental performance (does better environmental behavior lead to better environmental performance?). Lastly, as control factors, our analysis includes the explanatory factors identified in the appendix of Chapter 6: chemical manufacturing subsector, EPA region, facility size, and calendar year.

Readers less interested in the details of our statistical analysis than in the results of our statistical analysis may wish to skip to the beginning of Section 9.2.2, in which we interpret the results of our regression analysis.

To estimate the functional relationship between environmental performance and the identified multiple factors, we consider only the most simple functional relationship—a linear relationship—and employ only the most common type of multivariate regression analysis—ordinary least squares regression. Both choices are sufficient for our purposes.

We possess longitudinal or panel data on environmental performance since for each relevant facility we possess an observation for every relevant month of our sample period between January 1999 and March 2003. (As described in Chapter 4, all facilities do not face discharge limits for all regulated pollutants in all months of our sample period; thus, we must apply the qualifier "relevant" to our depiction of the panel data set.) In order to exploit and accommodate this panel data structure, we employ a fixed-effects estimator, which incorporates a set of facility-specific indicator factors: one indicator for each facility. These factors control for heterogeneity (that is, characteristic differences) across the facilities that is fixed over time for each facility. This approach seems appropriate when each facility is different in some way that is not captured by the other control factors and this difference does not change over time, as explained in Chapter 8.

Use of a fixed-effects estimator allows us to explore the effects of only behavioral forms that vary over time—count of audits and number of environmental employees—on environmental performance, as described further later in this chapter. Due to this limited use of the fixed-effects estimator, for the purposes of assessing the effects of *all* environmental behavioral forms on environmental performance, we also attempt to employ a random-effects estimator, which is also appropriate for exploiting and accommodating the panel data structure. This alternative estimator allows for heterogeneity across the facilities that relates to what are called *error terms* within the multivariate regression analysis. In general, these error terms represent elements that are not known to the researcher but that still influence the dependent variable. This approach seems appropriate when each facility is different in some way but the particular element is not known to the researcher.[7] In general, we do not interpret the results generated by the random-effects estimator since the results appear problematic in most cases.[8] Our attempt to explore the effects of all behavioral forms on performance is described in Section 9.2.2.3.

Lastly, we describe our ability to incorporate the many dimensions of deterrence into our multivariate analysis based on the actual use of the various government interventions: inspections, informal enforcement, and so on. Clearly, we are not able to incorporate a deterrence factor whenever the underlying government intervention is not performed in our sample or during our sample period, in the case of general deterrence. The full sample period runs from January 1999 to March 2003. Over this time period, regardless of the discharge form, our sample excludes the use of civil fines against all regulated chemical manufacturing facilities; thus, we are able to assess neither specific deterrence nor general deterrence. When estimating concentration-based TSS discharges, our sample also excludes the use of administrative formal nonsanctions and administrative fines against facilities in our sample (that is, no specific deterrence measures). In addition, when estimating concentration-based BOD discharges, our sample further excludes the use of informal enforcement actions and civil formal nonsanctions against facilities in our sample. Our analysis excludes additional deterrence measures when we explore subsample periods within the full sample period of January 1999 to March 2003, as described later in this chapter.

9.2.2. Identifying Subsamples and Interpreting Estimation Results

With this framing in mind, we identify the subsamples used to estimate the functional relationships involving environmental performance and interpret the estimation results for each subsample. Based on our interpretations, we assess whether the results provide evidence to support our hypothesis that greater deterrence prompts better environmental performance.

Our use of multivariate regression analysis generates many empirical results, including estimates of coefficient magnitudes and the associated standard errors. The coefficient magnitudes represent the multiplicative factors applied to the independent variables within the estimated linear functional relationship between the dependent variable, environmental performance, and the independent variables, for example, deterrence. (For example, m represents the coefficient magnitude in the following functional relationship between a dependent variable y and a single independent variable x: $y = b + mx$.) The standard error associated with a coefficient reflects the "noisiness" of the coefficient magnitude and measures the "tightness" of the coefficient estimate. By comparing the coefficient magnitude to its standard error, the multivariate regression analysis is able to assess the statistical significance of each independent variable's coefficient estimate, as reflected in each coefficient's p-value. This assessment stems from a null hypothesis of a zero coefficient magnitude, which implies that a particular independent variable has no influence on the dependent variable. We deem a coefficient estimate statistically significant when our analysis is able to reject the null hypothesis of a zero coefficient at a significance level of 10 percent or less, as revealed by a p-value of 0.10 or less. We deem a coefficient estimate statistically insignificant when our analysis is not able to reject the null hypothesis of a zero coefficient at a significance level of 10 percent or less, as revealed by a p-value greater than 0.10.

We report and interpret the coefficients relating to the influence of deterrence on environmental behavior, along with the coefficients relating to discharge limits and environmental behavior. We use common notation to identify the coefficient magnitude (β) and the level of statistical significance (p).

9.2.2.1. Full Sample Period.   We explore eight forms of environmental performance. The subsamples used for our multivariate regression analysis differ between the two pollutants—TSS and BOD—and between the two measurement forms—quantity and concentration. (The subsamples do not differ between the two frames—absolute and relative.) Based on the full sample period—January 1999 to March 2003, the following performance-specific subsample sizes apply:

TSS quantity based: 3,253 observations

TSS concentration based: 1,266 observations

BOD quantity based: 2,899 observations

BOD concentration based: 860 observations

Clearly, the full sample spans a time period that complicates our ability to incorporate the multiple measures of environmental behaviors into our set of explanatory variables. For example, we measure audits and environmental employees for each calendar year between 1999 and 2001. We possess no data on these behaviors for the calendar years 2002 and 2003. We measure other forms of environmental behavior, such as audit team composition, only for the twelve-month period preceding the completion of our survey. Consequently, we possess no data on these behavioral forms prior to March 2001 for any survey respondent, since the first month of survey completion is March 2002. We measure other forms of behavior, such as external consultant expenditures, only for the three-year period preceding survey completion. Thus, we possess no data on these behavioral forms prior to March 1999.

Rather than contorting our data on environmental behavior to fit the full sample period, we utilize the full sample period to estimate the functional relationship between environmental performance and a set of factors that excludes measures of environmental behavior. This approach is justifiable for two reasons. First, our use of a fixed-effects estimator at least controls for heterogeneity across facilities that does not vary over time. Most likely environmental behavior varies over time for a particular facility over our fifty-one-month full sample period. Certainly, our analysis of audit counts and environmental employment levels demonstrates that these dimensions vary over time for much of our sample. This point notwithstanding, if the general nature of environmental behavior does not vary much over time, then our use of a fixed-effects estimator helps to control for variation in environmental behavior across facilities (but not over time for any given facility). Second, by excluding environmental behavioral measures from our set of explanatory factors, our multivariate analysis may be able to assess jointly the direct and indirect effects of government interventions on environmental performance by interpreting the estimated relationship between performance and interventions. The indirect effect involves two links: one link from government interventions to environmental behavior and a second link from behavior to environmental performance. Chapter 8 examines the first link and Chapter 6 examines the second link. If we included environmental behavioral measures in our explanatory factor set, the multivariate analysis would be able to isolate the direct effect of government interventions on performance. In this case, the analysis would control for variation in environmental behavior prompted by government interventions. However, without the inclusion of environmental behavioral measures in the explanatory factor set, the multivariate analysis allows the prompted variation in environmental behavior to "creep" into the estimated relationship between performance and interventions.[9] (Note that this interpretation of our estimation results does not rely on any assumption that greater deterrence from government interventions leads to better environmental behavior.)

Given this perspective, we interpret the estimation results for each form of environmental performance in turn. First, we consider TSS discharges, starting with quantity-based discharges. Multivariate regression results reveal that absolute TSS quantity-based discharges appear to depend positively on both federal inspection–related specific deterrence and state inspection–related general deterrence and negatively on informal enforcement–related general deterrence (respectively, $\beta = 110$, $\beta = 91.9$, $\beta = -189$; $p = 0.000$, $p = 0.000$, $p = 0.024$), without adjustment for production. With production adjustment, absolute TSS quantity-based discharges depend negatively on federal inspection–related general deterrence ($\beta = -1.55$, $p = 0.076$). Thus, our results are not robust to the adjustment for production. The latter estimate set provides more evidence to support our hypothesis: increased general deterrence from state inspections leads to better environmental performance (that is, lower absolute quantity-based TSS discharges). Multivariate results also reveal that quantity-based TSS relative discharges depend negatively on both informal enforcement–related specific deterrence and federal inspection–related general deterrence, regardless of production adjustment (when adjusted: respectively, $\beta = -0.0152$, $\beta = -0.440$; $p = 0.011$, $p = 0.032$). Both results provide evidence to support our hypothesis: greater deterrence from informal enforcement and federal inspections improves environmental performance (that is, lowers TSS relative discharges).

Considering concentration-based discharges, multivariate results reveal that TSS absolute discharges depend on no deterrence measures, while TSS relative discharges depend negatively on federal inspection–related specific deterrence and civil formal nonsanction–related general deterrence (respectively, $\beta = -0.102$, $\beta = -4.67$; $p = 0.068$, $p = 0.019$). The latter results provide evidence to support our hypothesis: greater deterrence from federal inspections and formal nonsanctions improves TSS-related environmental performance.

Second, we consider BOD discharges, starting again with quantity-based discharges. Multivariate results reveal that absolute BOD quantity-based discharges appear to depend positively on both state inspection–related specific deterrence and general deterrence and negatively on informal enforcement–related general deterrence (respectively, $\beta = 7.25$, $\beta = 54.3$, $\beta = -101$; $p = 0.012$, $p = 0.000$, $p = 0.015$), without adjustment for production. With production adjustment, absolute BOD quantity-based discharges depend positively on state inspection–related general deterrence yet negatively on state inspection–related specific deterrence (respectively, $\beta = 1.69$, $\beta = -0.597$; $p = 0.092$, $p = 0.020$). Thus, the effect of state inspection–related general deterrence is robust to production adjustment, yet the effect of state inspection–related specific deterrence is clearly sensitive to the treatment of production. The effect of informal enforcement is also sensitive to this treatment. Regardless of the treatment, the evidence to support our hypothesis is mixed: greater deterrence may improve or undermine BOD-related environmental performance (that is, lower or higher BOD absolute discharges).

Multivariate results also reveal that quantity-based BOD relative discharges appear to depend positively on state inspection–related general deterrence, administrative formal nonsanction–related general deterrence, and administrative fine–related general deterrence, yet negatively on federal inspection–related general deterrence and informal enforcement–related general deterrence, regardless of production adjustment (when adjusted: respectively, $\beta = 0.0548$, $\beta = 0.267$, $\beta = 7.12\text{e-}5$, $\beta = -0.425$, $\beta = -0.197$; $p = 0.028$, $p = 0.015$, $p = 0.013$, $p = 0.092$, $p = 0.035$). These results provide mixed evidence to support our hypothesis: greater deterrence may improve or undermine BOD-related environmental performance.

Considering concentration-based discharges, multivariate results reveal that BOD absolute discharges appear to depend positively on federal inspection–related general deterrence, state inspection–related general deterrence, and administrative formal nonsanction–related general deterrence yet negatively on civil formal nonsanction–related general deterrence and administrative fine–related general deterrence (respectively, $\beta = 38.6$, $\beta = 0.711$, $\beta = 60.3$, $\beta = -59.3$, $\beta = -0.194$; $p = 0.001$, $p = 0.074$, $p = 0.004$, $p = 0.011$, $p = 0.022$). These results indicate that greater general deterrence from inspections leads to worse BOD-related performance, while greater general deterrence from fines leads to better BOD-related performance.

Greater general deterrence from formal nonsanctions improves or undermines BOD-related performance, depending on the source of the enforcement action: civil court or administrative court, respectively. Overall, the evidence to support our hypothesis is mixed.

Lastly, multivariate results reveal that BOD concentration-based relative discharges appear to depend positively on federal inspection–related general deterrence but negatively on both civil formal nonsanction–related general deterrence and administrative fine–related general deterrence (respectively, $\beta = 1.46$, $\beta = -2.82$, $\beta = -0.00702$; $p = 0.002$, $p = 0.003$, $p = 0.039$), which are dependencies applying also to BOD concentration-based absolute discharges. These results imply that greater general deterrence from inspections leads to worse BOD-related performance, while greater general deterrence from enforcement leads to better BOD-related performance.

Overall, these results provide only limited evidence in support of our general hypothesis that greater deterrence improves environmental performance. In general, the results reveal that greater deterrence fails to influence environmental performance. In certain cases, greater deterrence appears to undermine performance.

As a complement to our assessment of the link from imposed discharge limit level to environmental performance provided in Chapter 6, we interpret the multivariate results in order to explore the estimated relationship between environmental performance and the factor of discharge limit. Considering TSS discharges, quantity-based limits affect both absolute discharges and relative discharges. After adjusting both limits and absolute discharges for variation in production across facilities, the TSS quantity-based limit negatively affects TSS absolute discharges ($\beta = -0.233$, $p = 0.000$). Without any such adjustment, the effect proves statistically insignificant: $p = 0.571$. The TSS quantity-based limit also negatively affects TSS relative discharges ($\beta = -2.99E-4$, $p = 0.000$), as long as limits are not adjusted for variation in production. After adjusting limits for variation in production, the effect proves statistically insignificant: $p = 0.457$. The first result indicates that a decrease in a TSS quantity-based limit appears to prompt TSS absolute discharges to rise, which implies that TSS quantity-based limits are not merely ineffective but counterproductive. This result is quite surprising. The second result indicates that a decrease in a TSS quantity-based limit appears to prompt TSS relative discharges to rise, which implies that TSS quantity-based limits are binding. (The term *binding* is described in Chapter 6.)

At first blush, these two results are difficult to reconcile: if TSS quantity-based limits are counterproductive, then how can they be binding? Upon further reflection, we suggest that the first result is most likely driving the second result. If a reduction in the limit level prompts an increase in absolute discharges, by construction, the level of relative discharges (that is, discharge ratio) must rise. Clearly, the counterproductive result relating to absolute discharges disrupts our ability to interpret the result relating to relative discharges as evidence of a binding limit. Therefore, the two results collapse into a single interpretation: tighter TSS quantity-based limits do not prompt reductions in wastewater discharges. (TSS concentration-based limits significantly influence neither absolute nor relative discharges.)

Considering BOD discharges, concentration-based limits positively influence absolute discharges ($\beta = 0.0857$, $p = 0.094$). This result indicates that a decrease in the limit appears to prompt a decrease in absolute discharges, which implies that BOD concentration-based limits are effective at controlling pollution. BOD concentration-based limits do not significantly influence relative discharges. Moreover, BOD quantity-based limits significantly influence neither absolute nor relative discharges.

Overall, the multivariate results indicate that most forms of limits are neither effective at controlling pollution nor binding.

9.2.2.2. Subsample Period: January 1999 to December 2001. The full sample period does not facilitate our inclusion of environmental behavioral measures into the set of explanatory variables. To remedy this omission, we additionally consider subsample periods that facilitate this inclusion. The most obvious subsample period runs from January 1999 to December 2001. This subsample period cleanly facilitates the inclusion of behavioral forms measured for each calendar year between 1999 and 2001: audit count and environmental employment level. (Separately, we include the number of environmental employees in absolute terms and relative to all facility-level environmental employees; the two measures generate nearly identical estimates; later in this chapter we report only the estimates generated by including the absolute number of environmental employees.) By incorporating both of these behavioral forms into our explanatory factor set, we are able to isolate the direct effect of government interventions on environmental performance and the indirect effect of interventions on performance via the influence of interventions on environmental behavior. In order to assess the indirect effect, we first assess the effect of government interventions on audit count and environmental employment, as expressed in the results from Chapter 8, and second assess the effect of audit count and environmental employment on environmental performance, as expressed in the results of this chapter.

Based on the subsample period between January 1999 and December 2001, the following performance-specific subsample sizes apply:

TSS quantity based: 2,027 observations

TSS concentration based: 708 observations

BOD quantity ased: 1,833 observations

BOD concentration based: 486 observations

Given the smaller sample window (January 1999 to December 2001), when estimating TSS concentration-based discharges, relative to the full sample period, the sample additionally excludes the use of civil formal non-sanctions against specific facilities in the sample (that is, no specific deterrence). When estimating BOD concentration-based discharges, relative to the full sample period, the sample additionally excludes the use of administrative formal nonsanctions against other chemical manufacturing facilities located in EPA regions in which surveyed facilities reside (that is, no general deterrence).

Initially, we assess the direct effects of government interventions on environmental performance, interpreting the estimation results for each form of performance in turn. First, we consider TSS discharges, starting with quantity-based discharges. Similar to the full sample period, multivariate results reveal that absolute TSS quantity-based discharges appear to depend positively on both federal inspection–related specific deterrence and state inspection–related general deterrence and negatively on informal enforcement–related general deterrence (respectively, $\beta = 176$, $\beta = 199$, $\beta = -252$; $p = 0.000$, $p = 0.000$, $p = 0.023$), without adjustment for production, yet with production adjustment, absolute TSS quantity-based discharges depend negatively on state inspection–related general deterrence ($\beta = -2.67$, $p = 0.078$). Again, the latter estimate set provides more evidence to support our hypothesis: increased general deterrence from state inspections leads to better TSS-related environmental performance. Multivariate results also reveal that quantity-based TSS relative discharges appear to depend positively on informal enforcement–related general deterrence and negatively on state inspection–related specific deterrence, informal enforcement–related specific deterrence, and federal inspection–related general deterrence, regardless of production adjustment (when adjusted: respectively, $\beta = 0.174$, $\beta = -0.0146$, $\beta = -0.0212$, $\beta = -0.859$; $p = 0.070$, $p = 0.097$, $p = 0.011$, $p = 0.027$). Most of these results provide evidence to support our hypothesis: greater specific deterrence from inspections and enforcement and greater general deterrence from inspections improve TSS-related environmental performance. Considering concentration-based discharges, multivariate analysis reveals that both TSS absolute and relative discharges depend negatively on state inspection–related general deterrence (respectively, $\beta=-3.37$, $\beta=-0.101$; $p=0.084$, $p=0.036$). Both results provide evidence to support our hypothesis: greater general deterrence from state inspections improves TSS-related environmental performance.

Second, we consider BOD discharges, starting again with quantity-based discharges. Mostly similar to the full sample period, multivariate results reveal that absolute BOD quantity-based discharges appear to depend positively on both state inspection–related specific deterrence and general deterrence, along with federal inspection–related general deterrence, yet negatively on informal enforcement–related general deterrence (respectively, $\beta=13.3$, $\beta=355$, $\beta=153$, $\beta=-119$; $p=0.004$, $p=0.078$, $p=0.000$, $p=0.029$), without adjustment for production. Similar to the full sample period, with production adjustment, absolute BOD quantity-based discharges depend positively on state inspection–related general deterrence yet negatively on state inspection–related specific deterrence (respectively, $\beta=4.69$, $\beta=-0.650$; $p=0.007$, $p=0.077$). Regardless of the adjustment for production, the evidence to support our hypothesis is mixed: greater deterrence may improve or undermine BOD-related environmental performance. Multivariate results also reveal that quantity-based BOD relative discharges appear to depend positively on both administrative formal nonsanction–related general deterrence and administrative fine–related general deterrence, yet depend negatively on federal inspection–related general deterrence and informal enforcement–related general deterrence, regardless of production adjustment (when adjusted: respectively, $\beta=0.352$, $\beta=6.53\mathrm{E}{-}5$, $\beta=-0.651$, $\beta=-0.170$; $p=0.007$, $p=0.047$, $p=0.070$, $p=0.067$). As shown in the full sample period, these same results provide mixed evidence to support our hypothesis: greater deterrence may improve or undermine BOD-related environmental performance.

Considering concentration-based discharges, multivariate results reveal that both BOD absolute and relative discharges appear to depend positively on federal inspection–related general deterrence yet depend negatively on civil formal nonsanction–related general deterrence and administrative fine–related general deterrence (based on absolute discharges: respectively, $\beta=39.1$, $\beta=-104$, $\beta=-0.180$; $p=0.004$, $p=0.000$, $p=0.087$; based on relative discharges: respectively, $\beta=2.36$, $\beta=-4.92$, $\beta=-0.0120$; $p=0.000$, $p=0.000$, $p=0.014$). As shown in the full sample period, these same results indicate that greater general deterrence from inspections undermines BOD-related performance, while greater general deterrence from enforcement improves BOD-related performance.

Overall, these results provide only limited evidence to support our hypothesis that greater deterrence improves environmental performance.

In general, the results reveal that greater deterrence fails to influence environmental performance. In certain cases, greater deterrence appears to undermine performance.

A comparison of these results and the full sample period results reveals that most conclusions are robust to the choice of sample period and the incorporation of two key behavioral measures into our multivariate regression analysis as explanatory factors, that is, independent variables.

We next assess the indirect effects of government interventions on environmental performance. For this assessment, we first review the estimated effects of government interventions on environmental employment level and audit count, as presented in Section 8.2.2.2 of Chapter 8. In the case of environmental employees, the relevant multivariate results indicate that civil formal nonsanction specific deterrence negatively influences the absolute number of environmental employees. Otherwise, deterrence does not influence the absolute number of environmental employees. The multivariate results also indicate that state inspection–related general deterrence and civil formal nonsanction–related general deterrence positively influence the relative number of environmental employees, while administrative fine–related specific deterrence and general deterrence both negatively influence the relative number of environmental employees. Given the greater prevalence of influential factors, we choose to interpret the effect of deterrence on the relative number of environmental employees. In the case of audits, the relevant multivariate results indicate that civil formal nonsanction–based specific deterrence appears to influence positively the annual count of audits. Otherwise, deterrence does not influence the count of audits.

As the second component in our exploration of indirect effects, we assess the influence of environmental employment level and audit count on the various forms of environmental performance. Based on the multivariate analysis of environmental performance, the results indicate that neither type of environmental behavior influences TSS discharges, regardless of the measurement form and frame. While TSS discharges always negatively depend on the number of environmental employees and nearly always depend negatively on the count of audits (the exception is absolute concentration-based discharges), indicating that better behavior leads to better performance, none of these dependencies are statistically significant (every p-value equals or exceeds 0.232). In contrast, BOD discharges significantly depend on the number of environmental employees in two cases. When environmental employees are related to the number of overall facility-level employees, both quantity-based (production-adjusted) absolute discharges and concentration-based relative discharges depend negatively on the level of environmental employment (respectively, $\beta = -78.7$, $\beta = -4.86$;

p=0.102, p=0.010). Quantity-based (production-adjusted) BOD absolute discharges significantly depend also on the absolute number of environmental employees ($\beta = -0.052$, p=0.004).

By considering jointly these two sets of multivariate results, we identify
the indirect effects of government interventions on environmental performance. In the case of TSS discharges, no indirect effects exist. However, in
the case of BOD discharges, four indirect effects exist. By influencing
the relative number of environmental employees, greater state inspection–
related general deterrence and greater civil formal nonsanction–related general deterrence indirectly improve both quantity-based absolute discharges
and concentration-based relative discharges. In contrast, greater administrative fine–related specific deterrence and greater general deterrence indirectly
undermine both quantity-based absolute discharges and concentration-
based relative discharges. These results appear to indicate that less severe
deterrence indirectly improves BOD-related environmental performance
yet the most severe form of deterrence—fines—indirectly undermines environmental performance.

9.2.2.3. Subsample Period: Twelve-Month Period Preceding Survey.
The subsample period between January 1999 and December 2001 facilitates the inclusion of only two behavioral forms: audit count and environmental employment level. In order to incorporate all behavioral
forms into our explanatory factor set, we consider the twelve-month period preceding the completion of our survey along with the month of
survey completion, which differs across survey respondents. For each
survey respondent, the full thirteen-month period lies somewhere between March 2001 and March 2003. Based on this respondent-specific
time period, the following performance-specific subsample sizes apply:

TSS quantity based: 581 observations

TSS concentration based: 230 observations

BOD quantity based: 480 observations

BOD concentration based: 126 observations

Given these smaller sample windows, when estimating either TSS or BOD
quantity-based discharges, relative to the full sample period, the sample
additionally excludes the use of administrative formal nonsanctions and
administrative fines against specific facilities in the sample. When estimating TSS concentration-based discharges, relative to the full sample period,
the sample additionally excludes the use of both administrative and civil
formal nonsanctions and administrative fines against specific facilities in
the sample. When estimating BOD concentration-based discharges, relative
to the full sample period, the sample additionally excludes the use of fed-

eral inspections, informal enforcement actions, both administrative and civil formal nonsanctions, and administrative fines against specific facilities in the sample and the use of civil formal nonsanctions against other chemical manufacturing facilities located in EPA regions where the surveyed facilities reside (that is, no general deterrence).

We incorporate all of the behavioral forms into our multivariate regression analysis as explanatory factors, that is, independent variables. However, the fit depends on the measurement of the behavioral form. We easily incorporate the behavioral forms measured over the preceding twelve-month or preceding three-year period. In the case of behavioral forms measured for each calendar year between 1999 and 2001, we must stretch the relevance of these measures beyond the identified three-year window. In particular, we assume that the count of audits and number of environmental employees reported for 2001 sufficiently applies to those months lying within the calendar years 2002 and 2003, which represents the strong majority of the subsample. (This assumption does not prove problematic for identifying the effects of deterrence given our data structure and choice of estimator, as explained in the next paragraph.)

Regardless of the behavioral form, our data structure implies that the measured level of behavior does not vary over time for any given facility. Consequently, our use of a fixed-effects estimator eliminates our ability to estimate the effect of each behavioral form on environmental performance. In essence, we capture the effect of environmental behavior, along with any other facility-specific factor that does not vary over time, by incorporating facility-specific indicators into our set of explanatory factors. Since the facility-specific indicators do not vary over time, no "explanatory room" remains for other factors that do not vary over time. In other words, the analysis is not able to discern between the effects of multiple factors that vary in an identical fashion (that is, they are perfectly correlated). Unlike the full sample period, over the respondent-specific thirteen-month window, environmental behavior may not be expected to vary meaningfully. If this is true, by controlling for the time-invariant effect of environmental behavior, we improve our ability to isolate the effect of deterrence on performance.

In contrast to the fixed-effects estimator, our use of a random-effects estimator retains our ability to estimate the effect of each behavioral form on environmental performance. Yet, as noted earlier, random-effects estimates may prove problematic. Our assessment of the random-effects estimates reveals that only the random-effects estimates involving TSS quantity-based discharges are worthy of interpretation.[10] Thus, for our exploration of deterrence, we continue to interpret only the fixed-effects estimates. (This focus does not seem problematic since, when worthy of interpretation, the random-effects estimates generate reasonably similar conclusions and the comparable fixed-effects estimates reveal a stronger relationship between environmental performance and deterrence factors.) In contrast, when exploring the effects of behavior on environmental performance, we exclusively interpret the random-effects estimates generated by estimating the relationship between TSS quantity-based discharges and an explanatory factor set that includes all behavioral forms.

With this focus, we first interpret the effects of deterrence on environmental performance. As with the full sample period and the three-year period of 1999 to 2001, the multivariate results reveal mixed evidence for our hypothesis that greater deterrence leads to better environmental performance. Consider first TSS discharges. TSS quantity-based absolute discharges depend positively on state inspection–related specific deterrence ($\beta = 4.04$, $p = 0.084$), contrary to our hypothesis. When adjusted for production, these absolute discharges depend positively instead on federal inspection–related general deterrence ($\beta = 79.4$, $p = 0.100$). As with results reported previously, the relationship between quantity-based absolute discharges and deterrence depends on whether the analysis adjusts for production. TSS quantity-based relative discharges depend positively on both administrative formal nonsanction–related general deterrence and administrative fine–related general deterrence yet depend negatively on both federal inspection–related specific deterrence and general deterrence, along with state inspection–related general deterrence and informal enforcement–related general deterrence, regardless of production adjustment, with a single exception in the case of informal enforcement–related general deterrence (when adjusted for production: respectively, $\beta = 3.04$, $\beta = 0.00700$, $\beta = -0.289$, $\beta = -1.84$, $\beta = -0.297$, $\beta = -1.05$; $p = 0.005$, $p = 0.008$, $p = 0.000$, $p = 0.027$, $p = 0.003$; $p = 0.044$). TSS concentration-based discharges do not depend on deterrence, with the exception of the negative relationship between TSS concentration-based absolute discharges and federal inspection–related general deterrence ($\beta = -68.5$, $p = 0.098$).

The multivariate results relating to BOD discharges reveal even fewer links from government interventions to environmental performance, though the estimated links generally support our hypothesis. When adjusted for production, BOD quantity-based absolute discharges depend negatively on state inspection–related specific deterrence, federal inspection–related general deterrence, and informal enforcement–related general deterrence (respectively, $\beta = -1.52$, $\beta = -90.2$, $\beta = -36.1$; $p = 0.043$, $p = 0.009$, $p = 0.102$). In this case, the demonstrated evidence exclusively supports the hypothesis that greater deterrence leads to better environmental performance; that is, no results reject the hypothesis. (Without the production adjustment, BOD quantity-based absolute discharges depend on only state inspection–

related specific deterrence: $\beta = -4.38$, $p = 0.061$.) BOD quantity-based relative discharges depend negatively on federal inspection–related general deterrence ($\beta = -1.36$, $p = 0.083$). Again the demonstrated evidence is exclusively supportive, but in this case extremely weak. (Even this single dependency disappears if the BOD limits are not adjusted for variation in production.) Both BOD concentration-based absolute discharges and relative discharges depend positively on administrative formal nonsanction–related general deterrence (respectively; $\beta = 242$, $\beta = 4.15$; $p = 0.000$, $p = 0.0300$), contrary to our hypothesis.

Finally, we explore the effects of environmental behavior on environmental performance. As noted previously, we exclusively interpret the random-effects estimates generated by estimating the relationship between TSS quantity-based discharges—both absolute and relative—and an explanatory factor set that includes all behavioral forms. These estimates reveal that TSS absolute discharges depend positively on audit team composition, managerial concern for noncompliance, and attendance at environmental training sessions (respectively, $\beta = 57.0$, $\beta = 74.5$, $\beta = 0.953$; $p = 0.036$, $p = 0.009$, $p = 0.043$) yet depend negatively on the usefulness of compliance assistance materials, BOD-related treatment technology, and audit count (respectively, $\beta = -38.2$, $\beta = -63.7$, $\beta = -1.81$; $p = 0.113$, $p = 0.002$, $p = 0.020$; the effect of usefulness of compliance assistance materials is nearly marginally significant). Increases in the second set of behavioral forms appear to improve environmental performance. For example, a better BOD-related treatment technology lowers TSS quantity-based absolute discharges. (Surprisingly, a better TSS-related treatment technology does not influence TSS discharges.) However, increases in the first set of behavioral forms seem to undermine environmental performance. For example, a greater quantity of environmental training, as measured by attendance, increases TSS quantity-based absolute discharges.

Multivariate results also reveal that TSS quantity-based relative discharges depend positively on requested assistance from regulators and days of environmental training (respectively, $\beta = 0.223$, $\beta = 0.00590$; $p = 0.066$, $p = 0.013$) yet depend negatively on monitoring protocol strength, usefulness of compliance assistance materials, and audit count (respectively, $\beta = -0.217$, $\beta = -0.191$, $\beta = -0.00575$; $p = 0.047$, $p = 0.087$, $p = 0.115$; the effect of audit count is nearly marginally significant). Again, the increases in the second set of behavioral forms appear to improve environmental performance. Regardless of the frame—either absolute or relative—more useful compliance assistance materials and more frequent audits both lower TSS quantity-based discharges. However, increases in the first set of behavioral forms seem to undermine environmental performance. For example, regardless of the frame—either absolute or relative—an increase in some component of environmental training quantity—either number of days or extent of attendance—raises TSS quantity-based discharges.

9.2.2.4. Subsample Period: Three-Year Period Preceding Survey. As yet another approach for exploring the effects of deterrence on environmental performance, we consider the three-year period preceding the completion of our survey along with the month of survey completion, which differs across survey respondents. For each survey respondent, the full thirty-seven-month period lies somewhere between March 1999 and March 2003. In order to incorporate data on audit count and environmental employment level within the recorded time frame, we end the subsample period in December 2001. So that each facility (potentially) offers the same number of observations to the subsample, we start the subsample period in March 2000. Thus, the chosen subsample period runs between March 2000 and December 2001. The following performance-specific subsample sizes apply:

TSS quantity based: 1,102 observations

TSS concentration based: 408 observations

BOD quantity based: 979 observations

BOD concentration based: 228 observations

These samples are smaller than the samples based on the full sample period. When estimating either TSS or BOD concentration-based discharges, the sample additionally excludes the use of administrative and civil formal nonsanctions and administrative fines against specific facilities in the sample. When estimating BOD concentration-based discharges, the sample even further excludes the use of informal enforcement actions against specific facilities in the sample and the use of informal enforcement actions and civil formal nonsanctions against other chemical manufacturing facilities located in EPA regions where the surveyed facilities reside (that is, no general deterrence).

Into our set of explanatory factors (that is, independent variables), we incorporate the behavioral forms measured over the preceding three-year period—compliance assistance materials' usefulness, request for regulatory assistance, external consulting expenditures, TSS-related treatment technology, and BOD-related treatment technology—and those behavioral forms measured for each calendar year between 1999 and 2001—count of audits and number of environmental employees. For the first set of behavioral forms, our data structure implies that the measured level of behavior does not vary over time for any given facility, thus eliminating our ability to identify the independent effect of each of these behavioral forms using a fixed-effects estimator. In essence, the fixed-effects estimator captures the effects of these environmental behavioral forms jointly, along with any other facility-specific factor that does not vary over time, by incorporating facility-specific indicators into our set of explanatory factors. Given this challenge, the general inconsistency of the relevant random-effects estimates, and our preceding efforts to assess the effects of all environmental behavioral forms on environmental performance, in this section we do not attempt to assess the effects of behavioral forms measured over the preceding three-year period on environmental performance.

Instead, we focus our efforts on answering the key research question: does greater deterrence from government interventions lead to better environmental performance? Based on multivariate regression analysis of the chosen subsample using the identified set of explanatory factors, the results provide limited support for a definitive answer of yes. Consider first TSS discharges. Regardless of the measurement form and regardless of the frame, TSS discharges negatively depend on state inspection–related general deterrence.[11] Across the board, greater general deterrence from state inspections improves TSS-related environmental performance. In addition, TSS quantity-based relative discharges depend negatively on federal inspection–related specific deterrence, regardless of production adjustment (when adjusted: $\beta = -0.0697$, $p = 0.051$). Thus, greater specific deterrence from federal inspections improves TSS-related environmental performance when measured by the extent of compliance.

Consider next BOD discharges. Similar to TSS discharges, BOD quantity-related discharges—both absolute and relative—depend negatively on state inspection–related general deterrence.[12] In addition, BOD quantity-based relative discharges depend negatively on administrative formal nonsanction–related general deterrence, regardless of production adjustment (when adjusted: $\beta = -2.32$, $p = 0.082$). These two sets of results imply that greater general deterrence from both inspections and enforcement improves BOD-related environmental performance.

Overall, these results indicate that greater deterrence from both federal and state inspections and enforcement improves environmental performance. In particular, general deterrence from state inspections improves environmental performance regardless of the pollutant regulated. While several deterrence factors do not influence environmental performance, none appear to undermine environmental performance. Of all the considered sample or subsample periods, this subsample period provides the cleanest depiction of the functional relationship between environmental performance and deterrence relating to government interventions. While we might wish to highlight this clean depiction, its unique character only proves to emphasize that our results are clearly not robust to the choice of sample/subsample period and are mixed for all other sample/subsample periods.

9.2.2.5. Subsample Period: January 1999 to June 2001. Finally, we consider the effect of injunctive relief and SEPs on environmental performance. As noted in Chapter 7, our data on injunctive relief and SEPs ends in June 2001. Consequently, in order to incorporate deterrence measures relating to injunctions and SEPs into our set of explanatory factors, we limit the sample period to the months between January 1999 and June 2001, inclusively. Even though the explanatory factor set continues to include deterrence measures relating to inspections and other forms of enforcement, we interpret only the multivariate results relating to injunctions and SEPs.

Within the chosen sample period, the most obvious behavioral forms to include in our explanatory factor set are the count of audits and the level of environmental employment, since these measures are recorded for each calendar year between 1999 and 2001. In general, we are able to incorporate behavioral forms recorded for the three-year period preceding survey completion since this time frame overlaps greatly with the chosen subsample period. However, our use of a fixed effects estimator eliminates this option since these behavioral measures do not vary over time. Thus, we are not able to identify their influence separate from the influence of the facility-specific indicators.

Based on the identified subsample period, the following performance-specific subsample sizes apply:

TSS quantity based: 1,664 observations

TSS concentration based: 563 observations

BOD quantity based: 1,515 observations

BOD concentration based: 389 observations

For all of these subsamples, the data reveal the use of only certain dimensions of SEPs and injunctions. When estimating either TSS or BOD quantity-based discharges, the sample reveals the use of administrative injunctions against specific facilities in the sample, in addition to other chemical manufacturing facilities (that is, both specific and general deterrence), and the use of civil SEPs against other chemical manufacturing facilities but not facilities in the sample (that is, only general deterrence). When estimating TSS concentration-based discharges, the sample drops administrative injunction–related specific deterrence. When estimating

BOD concentration-based discharges, the sample additionally drops civil SEP–related general deterrence.

In general, the multivariate results indicate that neither SEPs nor injunctions influence environmental performance. As the single exception, TSS quantity-based relative discharges depend positively on civil SEP–related general deterrence yet negatively on administrative injunction-related general deterrence, regardless of whether the TSS quantity limit is adjusted for variation in production (based on production-adjusted results: respectively, $\beta = 0.00159$, $\beta = -7.34E-5$; $p = 0.104$, $p = 0.091$). These two results provide mixed evidence for our hypothesis; in this case, greater deterrence from nonfine sanctions may either improve or undermine environmental performance.

9.3. SUMMARY

As its primary purpose, this chapter inquires whether government interventions—inspections and enforcement actions—help to induce better environmental performance. Our effort to answer this question involves two empirical approaches: (1) a survey of facilities regarding their perceptions of the effectiveness of government interventions at prompting better compliance with NPDES discharge limits, and (2) multivariate regression analysis to estimate a functional relationship between actual discharges and government interventions. Our two empirical approaches appear to generate different conclusions. While the assessment of chemical manufacturing facilities' perceptions generally reveals that government interventions are believed to be effective, a majority of our multivariate regression results do not support this claim. Three broad reasons may explain this possible difference. First, the survey respondents may not have answered our survey questions honestly, distorting their responses to fit what they believed our research team and the EPA wished to hear. (When contacting potential respondents, we stated explicitly that our study was funded by the EPA.) If this distortion is true, the survey responses may not be reliable indicators of true perceptions. Second, the survey respondents may have answered our questions honestly but they were simply mistaken about the true effectiveness of government interventions in certain cases. Third, our multivariate regression analysis may be generating erroneous results. We assess this possibility in Chapter 10.

This chapter also further explores whether discharge limits influence environmental performance and whether better environmental behavior leads to better environmental performance. Regarding the effects of discharge limits, our results indicate that limits generally appear neither effective nor binding, the exceptional case being BOD concentration-based limits, which are effective. Regarding the effects of environmental behavior, our results indicate that some forms of environmental behavior appear to influence environmental performance positively, as expected, yet other forms negatively influence performance, while several forms do not influence environmental performance in either way.

CHAPTER TEN

Summary, Conclusions, Policy Implications, and Future Research

10.1. SUMMARY

This book attempts to examine polluters' efforts to comply with waste-water discharge limits, their success at compliance, and the role of government monitoring and enforcement in inducing stronger efforts and better compliance. For this examination, our book empirically explores the discharge limits imposed on U.S. chemical manufacturing facilities permitted to discharge wastewater pollutants directly into surface-water bodies and these facilities' actual discharges in light of these limits. Using data gathered by an original survey, we also explore a wide variety of environmental management practices that should help to control wastewater pollution (that is, environmental behavior). Lastly, we analyze the apparent links from increased use of government interventions—inspections and enforcement actions—to better environmental behavior and environmental performance. Our sample period runs from January 1999 to March 2003.

Based on our empirical analysis, we attempt to answer a variety of research questions. Some of our research questions are broad in scope. Our first broad research question concerns discharge limits. Simply put, what is the variation in Clean Water Act discharge limits that are imposed on permitted wastewater-discharging facilities embedded within the National Pollutant Discharge Elimination System program? Our second broad research question concerns the actions that wastewater-discharging facilities are taking to comply with these limits; hereafter we refer to these actions as "environmental behavior." We pose a simple question similar to our broad question concerning discharge limits: what is the environmental

behavior of permitted wastewater-discharging facilities? Our third broad
research question concerns the outcomes—in terms of discharges and
compliance—of these actions in the presence of the imposed discharge
limits; hereafter, we refer to these outcomes as "environmental performance." We ask this basic question, which is highly similar to our broad
question concerning environmental behavior: what is the environmental
performance of permitted wastewater-discharging facilities? Our fifth
broad research question concerns the actions taken by regulators to induce compliance with the imposed discharge limits. Our general question
is the following: what are regulators doing to induce environmental behavior that effectively controls discharges so that the discharges comply
with the imposed limits? Our specific question is the following: what
government interventions, namely, inspections and enforcement actions,
are regulators taking in order to induce compliance?

In addition to these broad research questions, we pose research questions that explore the relationships connecting discharge limits, environmental behavior, environmental performance, and government interventions. This additional set of research questions is more focused in scope.
Of these additional questions, the more important research questions focus on the effects of government interventions on environmental behavior
and environmental performance. In particular, we pose these two highly
policy-relevant questions: Do the identified government interventions
help to induce better environmental behavior? Do the identified government interventions help to induce better environmental performance? Other
questions focus on the effects of discharge limits on environmental behavior and environmental performance. Specifically, we pose these two policy-
relevant questions: How do the imposed discharge limits affect environmental behavior? How do the imposed discharge limits affect environmental
performance? The final research question focuses on the effect of environmental behavior on environmental performance: does better environmental behavior improve environmental performance?

10.2. CONCLUSIONS

Based on the results of our empirical analysis, which provide possible
answers to our research questions, we draw many conclusions. Consider
first our broad research questions. Based on the results generated by our
assessment of discharge limits, we conclude that quantity-based limits
vary greatly across facilities and to a much lesser extent over time, while
concentration-based limits vary somewhat across facilities and not
much at all over time. Based on the results generated by our assessment

of environmental behavior, we conclude that wastewater-discharging facilities engage in a variety of environmental management practices designed to facilitate compliance with discharge limits. Based on the results generated by our assessment of environmental performance, we conclude that facilities nearly always comply with their discharge limits and even strongly overcomply with their discharge limits a strong majority of the time, a result that is somewhat at odds with (and more comforting than) the findings of recent widespread noncompliance with discharge limits discussed in Section 1.1 of this book. Based on the results generated by our assessment of government interventions, we conclude that environmental protection agencies frequently monitor wastewater-discharging facilities, but not as much as they are required to do, and very infrequently take enforcement action against facilities. These findings are fully consistent with the portrait of less-than-vigorous enforcement action by Environmental Protection Agency (EPA) and state environmental agencies sketched out in the reports discussed in Section 1.1. Of course, the combination of very strong prevalence of compliance and highly infrequent enforcement is not surprising. Infrequent enforcement action would be far more troubling if our finding that compliance in the chemical manufacturing facility is not characteristic of other categories of industrial facilities subject to Clean Water Act discharge limits.

We next consider our research questions that explore the relationships across discharge limits, environmental behavior, environmental performance, and government interventions. Based on the results generated by our analysis of the links from discharge limits to environmental behavior, we conclude that the effect of discharge limits on environmental behavior varies across the various measures of environmental behavior explored. While tighter discharge limits prompt three forms of environmental behavior to improve, tighter discharge limits prompt four forms of environmental behavior to degrade. As an example of the former, tighter limits in the form of total suspended solids (TSS) concentration (but not in the forms of TSS quantity, biological oxygen demand [BOD] quantity, and BOD concentration) appear to prompt facilities to seek regulatory assistance. As an example of the latter, tighter limits in the forms of both TSS concentration and BOD concentration (but not in the forms of TSS quantity and BOD quantity) appear to induce facilities to hire fewer environmental employees. In most cases, limits do not seem to influence environmental behavior. As examples, neither TSS nor BOD limits seem to influence facilities' decisions to hire environmental employees of higher quality or to provide environmental training of greater quantity or quality.

Based on the results generated by our analysis of the effect of discharge limits on environmental performance, we conclude that both tighter TSS

discharge limits and tighter BOD discharge limits appear effective at reducing absolute discharges, but only tighter BOD discharge limits appear
binding in that facilities seem to struggle to comply with BOD limits. In
contrast, tighter TSS concentration–based discharge limits appear to induce facilities to discover cheaper means of controlling TSS discharges.

Based on the results generated by our analysis of the effect of environmental behavior on environmental performance, we conclude that better
environmental behavior, in most forms, appears to lead to better environmental performance. For example, an increase in the count of audits conducted annually improves environmental performance. In seven cases,
better environmental behavior may appear to lead to worse environmental performance. However, we analyze the link from environmental behavior to environmental performance using two analytical approaches:
two-sample tests and multivariate regression. By considering results from
both analyses, we identify evidence indicating that only one form of environmental behavior—the importance that management places on the role
of general plant employees in identifying and correcting conditions leading
to noncompliance—negatively influences environmental performance; that
is, greater importance leads to worse performance. In the remaining few
cases, the evidence linking better behavior to worse performance is only
partial. As an example, one of our two-sample test results indicates that
a stronger audit classification protocol appears to undermine environmental performance, yet the multivariate results indicate no statistically
significant link. As an example of the opposite combination, the multivariate results indicate that an increase in the compositional quality of a
facility's audit team appears to degrade environmental performance, yet
the two-sample test results indicate no statistically significant link. As an
even better example, multivariate results indicate that increased attendance
at environmental training sessions appears to degrade environmental performance, yet the two-sample test results indicate that increased attendance leads to better performance. For none of the examined forms of environmental behavior does our analysis fail to generate at least limited evidence
to support a link—positive or negative—from behavior to performance.

Lastly, we consider our research questions concerning the effect of
government interventions on environmental behavior and environmental
performance. Based on the results generated by our analysis linking government interventions to environmental behavior, we conclude that greater
deterrence stemming from government interventions in general does not
prompt better environmental behavior. In many cases, our results indicate that deterrence stemming from government interventions does not
influence facilities' behavioral decisions. As two examples, no form of
deterrence—neither specific deterrence nor general deterrence—stemming
from any type of intervention—neither inspection nor enforcement action—

appears to influence the compositional quality of a facility's audit team or the number of training days provided to environmental employees. As evidence against our hypothesis that greater deterrence prompts better behavior, in several cases, the greater use of government interventions appears to prompt worse environmental behavior. For example, increased general deterrence stemming from administrative fines appears to prompt facilities to hire fewer wastewater employees. In a minority of the cases, greater deterrence appears to prompt better behavior. As one example, greater specific deterrence stemming from civil formal nonsanctions appears to prompt a higher count of audits conducted on an annual basis. Overall, the influence of deterrence on behavior in general (considering all forms of behavior) clearly differs between the two forms of deterrence and across the various types of government interventions. Sometimes the influence even differs for a given form of environmental behavior. For example, our results indicate that increases in state inspection–related general deterrence and civil formal nonsanction–related general deterrence positively influence the number of environmental employees (relative to overall facility employees), yet increases in administrative fine-related specific deterrence and general deterrence both negatively influence the relative number of environmental employees.

Based on the results generated by our analysis linking government interventions to environmental performance, we conclude that greater deterrence stemming from government interventions in general does not prompt better environmental performance. In many cases, our results indicate that deterrence stemming from government interventions does not influence environmental performance. As evidence against our hypothesis that greater deterrence prompts better performance, in several cases, the greater use of government interventions appears to prompt worse environmental performance. In a minority of the cases, greater deterrence appears to prompt better performance. Overall, the influence of deterrence on environmental performance in general (considering all eight forms of performance: TSS versus BOD, quantity versus concentration, absolute discharges versus relative discharges) clearly differs between the two forms of deterrence and across the various types of government interventions. Sometimes the influence even differs for a given form of environmental performance.

10.3. POLICY IMPLICATIONS

The concern prompted by reports of lackluster industrial wastewater discharge performance and weak government enforcement that are discussed in Chapter 1 make this a particularly propitious time to take a close look at what kinds of environmental policies are best suited to achieving reductions in polluting discharges and concomitant improvements in the quality of aquatic resources. Unfortunately, we are unable to provide a "magic bullet" based on our study that provides an across-the-board prescription for improving the efficacy of the Clean Water Act as a pollution-control device. The breadth and variety of the questions we analyzed make it difficult to generate sweeping generalizations about the policy implications of our research. Instead, we provide in this book a series of findings that enforcement officials and other water-quality regulators may be able to use to craft effective strategies for addressing particular problems in the behavior and performance of regulated facilities in a variety of specific contexts. Further, as we indicate in the following section, our research generates additional questions about how facilities react to government actions such as discharge limits and enforcement. The further research we identify may usefully supplement the policy-based conclusions we are able to draw from this study.

Certain implications of our research seem relatively clear. The Clean Water Act initially sought to impose uniform discharge limits across all (or virtually all) facilities in a given industry as a means, among other things, of ensuring that facilities operating in one area or watershed did not experience competitive advantages or disadvantages as compared to facilities in the same industry subject to more or less stringent limits. Our study reveals that, at least among major facilities in the chemical manufacturing sector, uniformity is not the rule when it comes to quantity-based limits. As indicated above, concentration limits are more uniform. It may be that the different limits within this industrial sector result from the process by which the states are required to impose limit levels derived from concern over achieving ambient water-quality standards whenever these levels are more stringent than EPA's technology-based limit levels. If so, then the divergence in limits is consistent with the statutory design, notwithstanding an initial preference for industry-wide uniformity.

The data at our disposal did not permit us to determine the reason for the divergence in permitted discharge limits within the chemical manufacturing sector. If additional research reveals that the application of state water-quality standards is not driving the differences in quantity-based limits, then one way to engender the kind of uniformity in discharge limits that Congress sought to achieve when it adopted the Clean Water Act may be to use the form of limit—concentration based—that tends to be more alike than different across facilities within the industry. Continued reliance on quantity-based limits may produce the very competitive advantages and disadvantages that Congress sought to avoid.

Our findings concerning environmental management practices and compliance status are also revealing from a policy perspective. The variety of management practices in which chemical manufacturing dischargers engage as a means of complying with discharge limits indicates that there is no single route toward compliance. Regulators seeking to work with dischargers to facilitate compliance therefore have a considerable tool bag of management practices from which to choose in recommending changes in the practices engaged in by noncomplying facilities. If one method seems ill suited to the circumstances of a particular discharger, there may be others that would better fit the discharger's capabilities or preferences that provide the same opportunities to facilitate compliance. In addition, our empirical results indicate that, at least during the time period encompassed by our study, the industry we explore generated a strong compliance record. This record seems at odds with the perception, detailed in Chapter 1, that noncompliance with the Clean Water Act is rampant. If the chemical manufacturing sector is outperforming other industrial sectors, it may make sense for regulators to seek to replicate the management practices employed by the chemical manufacturing sector on a broader scale. It is also worth exploring whether EPA's identification of the chemical manufacturing industry as a priority sector for purposes of Clean Water Act compliance is somehow responsible for the apparently good track record we identified during the period of our study.

The results generated by our analysis of discharge limits suggest that a tightening of discharge limits generally is an ineffective means of improving environmental behavior. However, it is important to remember that the ultimate goal of tighter discharge limits is improvement of environmental quality, not changes in behavior that one would expect to yield lower levels of discharge. Accordingly, if, despite the absence of desirable changes in environmental behavior, actual discharges fall following the tightening of a discharge limit, the tightened limit will have achieved its ultimate purpose. Nevertheless, it is somewhat unsettling that tightened limits do not seem to prompt better environmental behavior, such as upgrades in treatment equipment, especially given that such upgrades seem to be an obvious way to comply with tightened limits.

The results generated by our analysis of discharge limits and environmental performance suggest, most fundamentally, that a tightening of discharge limits generally is an effective means of lowering discharges, as one would expect, especially given the norm of compliance in the chemical manufacturing sector during the period we study. Thus, the findings support the wisdom of the fundamental decision Congress made in adopting the strategy to move in stepwise fashion toward the Clean Water Act's nominal no-discharge goal through the imposition of increasingly stringent (technology-based) discharge limits.

The results also suggest, however, that not all limit reductions are created equal; that is, not all are likely to have the same kind of beneficial effects on actual discharges. In certain instances, facilities seem to comply easily with tighter limits, perhaps through relatively easily achieved changes in behavior. These easy fixes suggest that the previous limits were not too demanding and that further reductions in legally permissible discharges represent an effective way to achieve desired reductions in the volume of pollutants discharged as a means of reducing the impact of water pollution on health and the environment. In other cases (such as tighter TSS concentration–based limits), facilities respond to lower limits through a more complex process of seeking out newer, cheaper means of compliance; that is, the limits act as a forcing device to generate more effective or efficient control methods. In still other situations, facilities struggle to achieve tighter limits, perhaps because they are not willing to reduce profits through expenditures on new pollution-control techniques (for example, treatment technologies) or because those control techniques are not yet feasible. Given these different responses to tightened limits, regulators are advised to assess the likely discharge-reduction potential of different regulated industries. Efforts to force the development of better control techniques through limit reductions may be more successful in some contexts than in others.

The results from our analysis of environmental behavior and performance generally link better environmental behavior with better environmental performance, especially as reflected in Chapter 6. Thus, an intelligent strategy for environmental enforcement officials is to find ways to induce dischargers to improve their environmental behavior. Improvement in behavior, of course, is not the ultimate goal. Instead, what the government is most concerned about is performance—how much pollution regulated facilities are discharging, and, in particular, whether those facilities are complying with applicable discharge limits. Improvement in behavior is thus a means to an end, discharge reductions that will avoid or ameliorate the adverse impacts that water pollution has on health and the environment. A nuanced analysis of our findings suggests that certain forms of environmental behavior are more closely linked with better environmental performance than others. Those are the ones on which regulators should concentrate in making enforcement strategy decisions. Thus, for example, in crafting administrative injunctive relief, formulating SEPs, or settling judicially filed enforcement actions, it may make more sense for enforcement officials to extract from regulated facilities a commitment to increase the count of audits conducted annually than to demand a stronger environmental audit protocol. At least in certain contexts, frequency appears to be more important than the scope or content of the audit. Enforcement officials over time may discover that the kinds of behavioral changes that prompt better environmental performance are consistent across industries. Alternatively, they may find that differing industry customs and practices yield different impacts on performance from the same behavioral change, depending on the industry involved.

Our attempts to answer two research questions concerning the effects of government interventions on environmental behavior and performance generate somewhat confounding results. In many instances, greater deterrence stemming from government interventions does not seem to prompt either better environmental behavior or better environmental performance. We provide several possible explanations for these results in Chapters 8 and 9. These explanations include our small sample size, which leaves our analysis more vulnerable to the influence of outliers. Of course, in some instances, we do find that greater deterrence stemming from interventions does prompt improvements in particular forms of environmental behavior or environmental performance. In these instances, enforcement officials should take heart that their efforts to bring regulated facilities into compliance are apt to produce better behavior and performance either from the particular enforcement targets or from larger groups of facilities. In the instances in which those links are absent, one possible conclusion is that government interventions of the kind we assess are not as effective (or, at least, were not as effective for this particular group of facilities at this particular time) as the standard deterrence model would lead one to conclude. Under this reading of our results, policy makers might be inclined to shift their efforts from deterrence-based enforcement to nonadversarial, cooperative efforts to induce industry to adopt desirable changes in environmental behavior and to improve their environmental performance. In the case of changes in behavior, this approach would make the most sense in circumstances in which we detect a strong link between behavior and performance.

An alternative explanation, however, is that the enforcement actions taken were too weak to have the intended deterrent effects. This conclusion is certainly consistent with the findings in Chapter 1 concerning the infrequency and leniency that seem to have characterized the recent water-pollution enforcement landscape, at both the state and federal levels. The lesson under this interpretation of our findings is not that government interventions are ineffective, if not futile. Instead, the appropriate lesson would be that government needs to initiate enforcement actions more frequently and impose more onerous consequences when those actions succeed. After all, a rational polluter will choose the course of action likely to generate the least expense in environmental compliance. If federal and state officials seek monetary fines that do not manage (even when the government succeeds) to extract from violating facilities their avoided costs of compliance, then it will have been cheaper for the facilities concerned to exceed their discharge limits and pay the fines assessed against them than to have made the changes in behavior (including treatment facility upgrades) that would have been necessary to achieve compliance in the first place. Under those circumstances, it is little wonder that interventions have little or no beneficial impact on behavior or performance. Since our analysis controls for the severity of sanctions, such as fines, this particular explanation is meaningful only if facilities' responses to sanctions are nonlinear so that the response to an additional dollar of sanction rises as the size of the sanction rises. As the most extreme possibility, facilities may not respond at all to sanctions below a certain size threshold, waiting until the sanction "gets big enough" to warrant a response.

10.4. FUTURE RESEARCH

We claim that our analysis seems to contribute to a better understanding of pollution limits, polluting facilities' efforts to comply with these limits, polluting facilities' success at complying with these limits, environmental protection agencies' efforts to induce compliance in the form of inspections and enforcement actions, and the effectiveness of these inspections and enforcement actions at lowering absolute discharges and improving the extent of compliance. However, our analysis clearly does not generate conclusive evidence for answering any of the research questions that we posed. Instead, our analysis identifies just as many new questions as it attempts to answer. In order to explore better the original questions and to prompt additional exploration of the new questions, we propose the following future research.

First, we propose a larger sample for study. In order to analyze environmental behavior, we needed to implement our own survey of permitted discharging facilities. By focusing on major facilities, on whom the EPA systematically records data on discharge limits and measurements, our sample diminished to only 97 facilities. When exploring facilities' environmental behavioral decisions, this sample size is quite limiting. When linking government interventions to environmental behavior, a strong majority of our empirical results clearly demonstrate that government interventions are either ineffective or counterproductive. Evidence of ineffectiveness is not surprising given the size of our sample. However, evidence of counterproductive effects is another matter. Yet, even in this regard, we admit that our small sample size makes our analysis vulnerable to outliers. Even a few unusual responses to government interventions may be driving our empirical results and distorting the signals that those results otherwise might be sending.

Second, similar to the size of the sample, future research should explore creative ways to address the infrequent use of federal inspections and highly infrequent use of some enforcement actions. Given this reality, any empirical analysis must rely on only a few cases in order to discern the effects of these kinds of interventions on environmental behavior and performance. A larger sample need not alleviate this concern as long as the frequency of use remains reasonably uniform across the universe of regulated polluting facilities. Rather than merely expanding the sample, future research should search for a context where and when enforcement actions were used more frequently. Even though the results of this future research need not generalize to other contexts, at least this type of future analysis would be better able to deliver a clearer picture of effectiveness.

Third, future research should explore more deeply the identified counterproductive effects of government interventions on both environmental behavior and environmental performance. In this deeper exploration, future research should consider the possible flaws of our empirical analysis. For example, consider our analysis of environmental behavior. We admit that our use of two-sample tests may be problematic since these tests do not control for other possibly confounding factors (for example, facility size). However, the multivariate analysis generates counterproductive results too. Of course, our multivariate analysis may be missing the control factors necessary to remove those confounding influences. Fortunately, our panel data analysis in the case of audits and number of environmental employees controls for anything that varies across the individual facilities but does not vary over time for any individual facility. Perhaps consistent with this advantage, we generate only "effective" or "ineffective" results in the case of audits. However, we continue to generate some counterproductive results for the number of environmental employees.

Fourth, future research should pay more attention to isolating the influence of specific deterrence. Our ability to generate "effective" results for specific deterrence is undermined by the fact that regulators penalize only violators and inspect violators more frequently, sometimes described as "targeting." Given this pattern, we need not be surprised to learn that specific deterrence appears to prompt worse behavior. In hindsight, we acknowledge that our own analysis should have tried harder to isolate the effect of specific deterrence. Unfortunately, these greater efforts demand more sophisticated statistical analyses than those employed here. Given the intended audience for this monograph, we steered clear of more sophisticated analytical methods. Future research should find a better compromise position.

Fifth, future research should pay more attention to identifying the influence of general deterrence. Unlike specific deterrence, general deterrence is not vulnerable to targeting. Nevertheless, our analysis generates counterproductive results even in the case of general deterrence. Our analysis generates these results when employing both two-sample tests and multivariate analysis, including multivariate analysis that controls for differences across individual facilities that are difficult to measure other than by crudely using facility-specific indicators.

In contrast to specific deterrence, where the effect of an actual intervention applies to a specific facility, general deterrence relies upon the perceptions of facilities about the threat of future interventions, based on the experiences of other facilities. Clearly, it is difficult to measure perceptions, although our survey results represent a modest step in that direction. Our empirical analysis attempts to measure perceptions using a simple approach: interventions against other similar facilities over the current calendar year. Future research could explore a variety of ways for measuring perceptions, such as these:

1. Interventions against others over the preceding twelve months (similar to our specific deterrence measures) plus the current month
2. Interventions against others over the upcoming twelve months plus the current month
3. Interventions against others over the preceding six months and upcoming six months plus the current month

The first measure assumes that facilities generate their perceptions based exclusively on backward-looking expectations. The second measure assumes that facilities generate their perceptions based exclusively on forward-looking expectations. The third measure assumes that facilities generate their perceptions based on a combination of backward-looking and forward-looking expectations.

To understand better the influence of general deterrence on environmental behavior and performance, it also may be useful to measure the degree to which individual regulated facilities are aware of interventions taken against others, because in the absence of such knowledge, industry perceptions about future interventions may not be very well-informed or useful.

As part of this exploration of perceptions relating to general deterrence, future research should improve the likelihood that an increase in interventions against other similar facilities truly translates into an increase in an individual facility's perceived threat of interventions against that individual facility in the near future. This improvement would serve to reduce the possibility that the individual facility perceives the increase in interventions against other facilities as an indication that the "storm has passed" without harm to the individual facility, that is, that the regulator was looking for easy targets and it has already found them.

Future research on both specific deterrence and general deterrence may reveal whether the absence of stronger links from government interventions to better environmental behavior and performance in the chemical manufacturing industry, which is reflected in some aspects of our study, is anomalous. Are the results different if a longer period of time is studied? Are they different for other industries subject to Clean Water Act discharge limits? If the answer to the latter question is affirmative, it would be useful to ascertain what aspects of the chemical manufacturing industry lead it to respond to government interventions in ways that differ from those of other regulated facilities.

Sixth, future research should assess more thoroughly the variety of elements that represent environmental behavior. While our analysis considers a wide variety, it certainly does not exhaust the possibilities. With a larger set of behavioral elements, future research could explore more deeply which elements of environmental behavior help to improve environmental performance. As noted in Section 10.2, our analysis generates results indicating that upward of seven behavioral elements may negatively influence environmental performance; that is, better behavior may undermine performance. While we argue that the evidence appears meaningful for only a single behavioral element and one capturing only the stated importance that management places on conditions leading to noncompliance, the existence of these counterproductive effects may explain why increased use of government interventions appears to prompt worse behavior in some forms. If some elements of environmental behavior truly do not help to improve environmental performance, then the use of interventions may prompt facilities to reassess the usefulness of these counterproductive elements of behavior and reallocate their resources away from these particular behavioral elements.

Notes

Chapter One

1. Any such differences are beyond the scope of this book. As we indicate in Chapter 4, Section 4.4, we believe that our choice to focus on discharges of two pollutants, biological oxygen demand (BOD) and total suspended solids (TSS), by the chemical industry also makes the results of our empirical analysis relevant to the operations and compliance practices of other industries regulated under the Clean Water Act. These two pollutants are two of the most common pollutants, they comprise the standard measure of organic pollutant content of water, BOD is considered by EPA to be the most damaging of the conventional pollutants, and BOD is widely used as an index of the harm caused by effluents in water. In addition, EPA has relied on some of the same treatment technologies used to control TSS and BOD to establish effluent limitation guidelines for toxic pollutants regulated under the Clean Water Act.

Chapter Two

1. The institute is currently named the Institute for Policy and Social Research.

2. In this first stage, we estimate a functional relationship between the decision of whether or not to respond to our survey, which represents the dependent variable, and a set of factors that should explain this decision (i.e., explanatory factors or independent variables). The response decision represents a dichotomous, qualitative dependent variable. In order to accommodate the qualitative nature of this dependent variable, we employ a probit estimation model. The interested reader is referred to Greene (2003) for an in-depth description of the probit estimator. The eager reader is encouraged to read Maddala (1983).

Chapter Three

1. Other studies also examine the effect of government interventions on environmental performance. Barla (2007) examines the influence of inspections on wastewater discharges from Quebec pulp and paper plants. Dasgupta et al.

(2001) examine the influence of inspections on wastewater discharges from Chinese manufacturing companies. Gangadharan (2006) examines the influence of inspections on Mexican plants' environmental compliance. Wang et al. (2003) examine the influence of inspections on the degree of Chinese companies' compliance with wastewater discharge limits. Lanoie, Thomas, and Fearnley (1998) examine the influence of enforcement on Canadian pulp and paper mills' wastewater discharges.

2. Other studies also examine the effect of environmental behavior on environmental performance. Gangadharan (2006) examines the influence of two components of environmental behavior—employee education and environmental training—on Mexican plants' environmental compliance. Wang et al. (2003) examine the influence of one aggregate form of environmental behavior—the ratio of pollution-control-related operating costs to total operating costs—on the degree of Chinese companies' compliance with wastewater discharge limits.

Chapter Four

1. *U.S. Code* 33 (2006) § 1251(a).
2. *U.S. Code* 33 (2006) § 1311(a).
3. *U.S. Code* 33 (2006) § 1342.
4. *U.S. Code* 33 (2006) § 1344.
5. *U.S. Code* 33 (2006) § 1342(a).
6. *U.S. Code* 33 (2006) § 1362(11).
7. *U.S. Code* 33 (2006) § 1318.
8. *U.S. Code* 33 (2006) § 1342(k).
9. *U.S. Code* 33 (2006) § 1342(b); *Code of Federal Regulations* 40 (2008) §§ 123.21–123.23.
10. *U.S. Code* 33 (2006) § 1342(b); *Code of Federal Regulations* 40 (2008) §§ 123.26–123.27.
11. *Code of Federal Regulations* 40 (2008) § 123.1(d)(1).
12. *U.S. Code* 33 (2006) § 1342(d)(4).
13. *U.S. Code* 33 (2006) § 1319.
14. *E.I. du Pont de Nemours & Co. v. Train*, 430 U.S. 112 (1977).
15. *U.S. Code* 33 (2006) § 1311(b)(1)(A). Under *U.S. Code* 33 (2006) § 1311(b)(1)(B), municipal sewage treatment plants are subject to a separate set of effluent limitations that reflect the application of secondary treatment technology.
16. *U.S. Code* 33 (2006) § 1311(b)(2)(E).
17. *U.S. Code* 33 (2006) § 1314(a)(4).
18. *U.S. Code* 33 (2006) § 1311(b)(2)(A). Under *U.S. Code* 33 (2006) § 1362(13), toxic pollutants are those "which, after discharge and upon exposure, ingestion, inhalation, or assimilation into any organism, either directly from the environment or indirectly by ingestion through food chains, will, on the basis of information available to [the EPA], cause death, disease, behavioral abnormalities, cancer, genetic mutations, physiological malfunctions (including malfunctions in reproduction) or physical deformations, in such organisms of their offspring."
19. *Code of Federal Regulations* 40 (2008) part 414.

20. *Code of Federal Regulations* 40 (2008) § 414.11(a).

21. *Code of Federal Regulations* 40 (2008) part 415.

22. When appropriate, a permit may specify a schedule of compliance that requires compliance as soon as possible but not later than the applicable deadline under the Clean Water Act. *Code of Federal Regulations* 40 (2008) § 122.47(a)(1).

23. *Code of Federal Regulations* 40 (2008) § 124.6(d).

24. *Code of Federal Regulations* 40 (2008) § 122.41(i).

25. *Code of Federal Regulations* 40 (2008) § 122.43(a).

26. *Code of Federal Regulations* 40 (2008) § 122.41(a).

27. *Code of Federal Regulations* 40 (2008) § 122.46(a).

28. *Code of Federal Regulations* 40 (2008) § 122.41(f).

29. *Code of Federal Regulations* 40 (2008) § 122.44(a)(1).

30. *U.S. Code* 33 (2006) § 1311(b)(1)(C).

31. The establishment and implementation of water-quality standards is a joint federal-state effort. *U.S. Code* 33 (2006) § 1313(c)(3)–(4); *Code of Federal Regulations* 40 (2008) § 131.5.

32. *U.S. Code* 33 (2006) § 1313(c)(1).

33. *U.S. Code* 33 (2006) § 1313(c)(2).

34. *Code of Federal Regulations* 40 (2208) § 131.6 describes the minimum requirements for water-quality standards.

35. *Code of Federal Regulations* 40 (2008) § 131.10.

36. *Code of Federal Regulations* 40 (2008) § 131.11(g)–(h).

37. *Code of Federal Regulations* 40 (2008) § 131.2 states that a water-quality standard designates a use to be made of the water and sets criteria to protect the use.

38. *U.S. Code* 33 (2006) § 1314(a)(1).

39. *U.S. Code* 33 (2006) § 1313(d).

40. *U.S. Code* 33 (2006) § 1313(d)(1)(C).

41. *Code of Federal Regulations* 40 (2008) § 131.13.

42. *U.S. Code* 33 (2006) § 1370(1).

43. *Code of Federal Regulations* 40 (2008) § 122.44(d)(5).

44. *Code of Federal Regulations* 40 (2008) § 124.51(b).

45. *U.S. Code* 33 (2006) § 1342(a)(10).

46. *Code of Federal Regulations* 40 (2008) § 125.3(a)(2)(i)(B), (a)(2)(ii)(B).

47. *Code of Federal Regulations* 40 (2008) § 122.45(a).

48. *Code of Federal Regulations* 40 (2008) § 122.45(b)(2)(i).

49. *Code of Federal Regulations* 40 (2008) § 122.45(f)(1).

50. Of the 188 restricted pollutants, we also identify the 15 most prevalent pollutants based on the number of facility-month observations when the pollutant faces an operative limit. Again, we disregard flow and pH range excursions. Clearly, the most prevalent regulated pollutant is TSS, and the second-most prevalent regulated pollutant is BOD. Of the 4,604 relevant observations, TSS and BOD limits apply to 3,714 and 3,117 observations, respectively. Fortunately, the two lists of top 15 pollutants strongly overlap. Only two differences exist. Methylene chloride is included only in the first list, while phosphorus is included only in the second list. Since nearly all of the information on behavior is recorded for a particular facility, rather than a single pollutant, we select the list based on facilities.

51. Throughout this book, we employ commonly used characters to represent statistical features. We use the Greek character ρ to denote a correlation magnitude, which is standard notation. We use p to represent the statistical level of significance, and N to represent the number of observations or sample size.

52. Despite the correlations between TSS/BOD and other important pollutants, we acknowledge that a large portion of the sampled facilities face a wide variety of operative effluent limits. On this point, we assess the number of pollutants covered by operative limits, calculated for each separate facility. Ten percent of the facilities face an operative limit at some point during the sample period for 22 or more pollutants, while 25 percent of the facilities face an operative limit for 11 or more pollutants. In contrast, only 12 percent of the facilities face an operative limit for 2 or fewer pollutants.

53. For those facilities reporting only a single year's data, we use that single data point to represent the facility's number of employees; for those facilities reporting data for two or three years, we use the average number of employees. Ninety-five of the 97 facilities provide information on the number of employees for at least one of the three years. All 73 facilities providing production information also provide employee information.

54. The facility-specific mean number of employees on average equals 357.1 employees ($N = 4,182$).

55. For the quantity limits, the three pairwise correlations lie at or above 0.999. For production-adjusted quantity limits, the correlations lie at or above 0.9999; the correlation relating mean- and median-aggregated limits equals 1, implying identical values. For concentration limits, the correlations lie at or above 0.991. For concentration-equivalent limits, the correlation lies at or above 0.992.

56. All BOD-relevant facilities provide at least one year's worth of data on employees. Based on the full BOD sample of 3,468 observations, the facility-specific average number of employees equals 388.7 on average, with a standard deviation of 322.2.

Chapter Five

1. Our survey also gathers information on the specific budgetary amounts devoted to training. However, only 54 of the 97 respondent facilities provide these detailed data. We doubt that these data are representative of the sample, so we do not report them.

2. Fourteen facilities identify their TSS treatment technology as "other." We classify these cases as possessing the least effective treatment.

3. Nine facilities identify their BOD treatment technology as "other." We classify these cases as possessing the least effective treatment.

4. The ranking of audit teams with only internal members relative to audit teams with only external members does not prove important for our subsequent empirical analysis relating discharge limits to environmental behavior and relating environmental behavior to environmental performance.

5. While 96 of the 97 responding facilities indicate whether or not they hired an external consultant, only 88 facilities provide detailed information on the ac-

tual amount of consultant expenditures; not surprisingly, facilities appear reluctant to report dollar values.

6. Given the manner in which we measure certain environmental management practices in multiple ordered categories, it would prove difficult to assess whether differences exist between subsamples. In essence, the analysis would need to demonstrate that the distribution over the multiple categories differs between subsamples. We avoid this difficulty by reformulating the measures into two broader ordered categories.

7. For the behavioral forms measured in each of three calendar years, the two TSS subsamples for production-adjusted quantity limits and concentration-equivalent limits are defined by these median values: 1.30 kg of pollutant per 1,000 kg of product and 30.0 mg/L, while the two BOD subsamples for production-adjusted quantity limits and concentration-equivalent limits are defined by these median values: 1.62 kg of pollutant per 1,000 kg of product and 22.3 mg/L. For the behavioral forms measured in the preceding twelve-month period, the two TSS subsamples for production-adjusted quantity limits and concentration-equivalent limits are defined by these median values: 1.29 kg of pollutant per 1,000 kg of product and 30.0 mg/L; while the two BOD subsamples for production-adjusted quantity limits and concentration-equivalent limits are defined by these median values: 1.04 kg of pollutant per 1,000 kg of product and 21.0 mg/L. For the behavioral forms measured in the preceding three-year period, the two TSS subsamples for production-adjusted quantity limits and concentration-equivalent limits are defined by these median values: 1.30 kg of pollutant per 1,000 kg of product and 30.0 mg/L; while the two BOD subsamples for production-adjusted quantity limits and concentration-equivalent limits are defined by these median values: 1.62 kg of pollutant per 1,000 kg of product and 21.0 mg/L. (Section 5A.4 of the appendix further describes the sample used to examine behavior for each time frame explored.)

8. This significance is based on the median two-sample test (p = 0.049) but not the Wilcoxon two-sample test.

Chapter Five Appendix

1. We do not employ a two-sample means t-test, which requires a normal distribution in both subsamples. In all cases, we reject the null hypothesis of normality in at least one of the two subsamples tested (p < 0.01), regardless of the normality test employed: Shapiro-Wilk, Kolmogorov-Smirnov, Cramer–von Mises, Anderson-Darling.

2. The respective sample sizes are 94, 95, and 283. The respective mean values are 320.0, 340.3, and 341.9.

3. We do not employ a two-sample means t-test, which requires a normal distribution in both subsamples. In all cases, we reject the null hypothesis of normality in both of the two subsamples tested (p < 0.01), regardless of the normality test employed: Shapiro-Wilk, Kolmogorov-Smirnov, Cramer–von Mises, Anderson-Darling.

Chapter Six

1. To avoid further the influence of outliers, when analyzing BOD quantity discharges, we exclude 17 observations in which the ratio of absolute discharges to permitted discharges exceeds 10, which we consider as an upper bound on a reasonable ratio. This adjustment does not affect to any meaningful degree our depiction of the distribution of environmental performance and affects nearly none of the hypothesis test statistics; it does meaningfully affect the calculation of average values.

2. As an alternative test, we attempt to employ a two-sample means t-test. This alternative test is valid only if environmental performance is normally distributed in each subsample. We assess this condition using four distributional tests: Shapiro-Wilk, Kolmogorov-Smirnov, Cramer–von Mises, and Anderson-Darling. These tests assess whether or not the null hypothesis of a normal distribution can be rejected. We apply the criterion that the null hypothesis is rejected when the p-value associated with a given test lies at or below 0.05 (that is, the null hypothesis is rejected at a significance level of 5 percent). Based on the calculated distributional test statistics, we reject the null hypothesis of normality in both of the two subsamples at the 1 percent level in all relevant cases. Thus, we do not report the results of this alternative test.

3. For the purposes of testing our two hypotheses, we assess the one-tailed p-value associated with these two tests.

4. We assess the one-tailed p-value.

5. The absolute magnitude of the correlation coefficient for this relationship is smaller than the relationship between quantity-based limits and quantity discharges; this comparison is expected since the former relationship connects an absolute level to a ratio where the latter relationship connects two absolute levels.

6. As an alternative test, we attempt to employ a two-sample means t-test. This alternative test is valid only if environmental performance is normally distributed in each subsample. We assess this condition using four distributional tests: Shapiro-Wilk, Kolmogorov-Smirnov, Cramer–von Mises, and Anderson-Darling. These tests assess whether or not the null hypothesis of a normal distribution can be rejected. We apply the criterion that the null hypothesis is rejected when the p-value associated with a given test lies at or below 0.05 (that is, the null hypothesis is rejected at a significance level of 5 percent). Based on the calculated distributional test statistics, we reject the null hypothesis of normality in both of the two subsamples at the 1 percent level in all cases except the case of TSS treatment technology absence or presence in the subsample of lesser environmental behavior (that is, absence of TSS treatment technology), where one distributional test rejects the null hypothesis of normality at a level between 1 percent and 5 percent and another distributional test rejects this null hypothesis at a level between 5 percent and 10 percent. Thus, we do not report the results of this alternative test.

7. For the purposes of testing our two hypotheses, we assess the statistical significance revealed by the one-tailed p-values of these two tests.

8. We assess the one-tailed p-values.

9. The ranking of the subsample median values does not coincide with the ranking of subsamples based on the Wilcoxon two-sample test and median two-sample test, which employ scoring methods for assessing whether the distribution of one subsample is located at higher levels of the variable in question than the distribution of the other subsample.

10. The ranking of the subsample median values does not coincide with the ranking of subsamples based on the Wilcoxon two-sample test and median two-sample test, which employ scoring methods for assessing whether the distribution of one subsample is located at higher levels of the variable in question than the distribution of the other subsample.

11. The two-sample means t-test is valid if environmental performance is normally distributed in each subsample. We assess this condition using three distributional tests: Kolmogorov-Smirnov, Cramer–von Mises, and Anderson-Darling, which assess whether or not the null hypothesis of a normal distribution can be rejected. All of the calculated test statistics reject this null hypothesis at the 1 percent significance level. Thus, we must interpret the two-sample means t-test results cautiously.

12. For the purposes of testing, we assess the statistical significance revealed by the tests' two-tailed p-values since we have no a priori expectation about which subsample of facilities might generate a higher or lower level of environmental performance.

13. Under the assumption of equal variances, $p = 0.968$, and under the assumption of unequal variances, $p = 0.969$; the latter is more appropriate since the test of variances rejects the null hypothesis of equal variances ($p < 0.0001$).

14. Under the assumption of equal variances, $p = 0.581$, and under the assumption of unequal variances, $p = 0.616$; the latter is more appropriate since the test of variances rejects the null hypothesis of equal variances ($p < 0.0001$).

Chapter Six Appendix

1. A two-sample means t-test is appropriate if environmental performance is normally distributed in both subsectors being compared. In all cases, we reject the null hypothesis of normality in each subsample tested ($p < 0.01$), regardless of the four normality tests that we employ: Shapiro-Wilk, Kolmogorov-Smirnov, Cramer–von Mises, Anderson-Darling.

2. A two-sample means t-test is appropriate if environmental performance is normally distributed in both subsamples being compared. In all cases, we reject the null hypothesis of normality in each subsample tested ($p < 0.01$), regardless of the four normality tests that we employ: Shapiro-Wilk, Kolmogorov-Smirnov, Cramer–von Mises, Anderson-Darling.

3. As supporting evidence, both correlations between environmental performance and facility size prove significantly negative; that is, as size increases, discharge ratios fall.

Chapter Seven

1. *U.S. Code* 33 (2006) § 1342(b)(2)(B), (7); *Code of Federal Regulations* 40 (2008) § 123.27.

2. For a more complete description of the Clean Water Act authority and processes, see Hunter and Waterman (1996).

3. See, for example, *Code of Federal Regulations* 40 (2008) § 125.124.

4. *U.S. Code* 33 (2006) § 1318(a).

5. *U.S. Code* 33 (2006) § 1318(c).

6. *U.S. Code* 33 (2006) § 1319.

7. *U.S. Code* 33 (2006) § 1319(d).

8. *U.S. Code* 33 (2006) § 1319(d).

9. *U.S. Code* 33 (2006) § 1319(g)(1).

10. There is nothing in the Clean Water Act that prevents EPA from calculating economic benefits from the first date of violation, even if that is more than five years before the complaint was filed, as long as the statutory maximum penalty exceeds the settlement penalty amount (EPA, 1995).

11. The PCS database also identifies industrial user and pretreatment inspections. These inspection types do not relate to our sample of survey respondent facilities because none of them are industrial users.

12. An assessment of the inspections performed at survey respondent facilities reveals a highly similar pattern. EPA mostly uses compliance evaluation inspections (20). It never uses reconnaissance inspections or toxic inspections. In between, EPA performs 4 audit inspections, 3 biomonitoring inspections, and 2 other inspections. Of particular interest, EPA performs only 3 sampling inspections, in contrast to the 29 nonsampling inspections. Similar to EPA, state agencies mostly use compliance evaluation inspections (280). However, in contrast to EPA, state agencies heavily use reconnaissance inspections (250), which represents the second-most frequently used type. At the low end, state agencies infrequently perform toxic inspections (14), similar to EPA, and other inspections (12). In between, state agencies perform 25 biomonitoring inspections and 12 audit inspections. Of particular interest, state agencies perform 153 sampling inspections, in contrast to the 593 nonsampling inspections. Thus, state agencies perform sampling inspections more frequently than do EPA regional offices as a proportion of overall inspections performed.

Chapter Eight

1. We also attempt to employ a two-sample means t-test, which is valid only if the behavioral measure is normally distributed in each subsample. We assess this condition using four distributional tests: Shapiro-Wilk, Kolmogorov-Smirnov, Cramer–von Mises, and Anderson-Darling. The calculated distributional test statistics reject the null hypothesis of normality in both subsamples in all cases, regardless of the test employed, with the exception of five subsamples involving enforcement-related specific deterrence, where the particular subsample size equals 5, and three subsamples involving injunction-related specific deterrence, where

the particular subsample size equals 1. In both cases, small sample sizes complicate the tests' ability to reject the null hypothesis of normality.

2. We also describe the division of our sample into two subsamples for the purposes of assessing whether the distribution of a behavioral measure differs between these two subsamples: facilities facing below-median deterrence and facilities facing above-median deterrence. This division is specific to the particular time frame used to measure a behavioral form: (1) each of three calendar years between 1999 and 2001, (2) the twelve-month period preceding survey completion, and (3) the three-year period preceding survey completion. For the first set of behavioral forms, the two inspection-related subsamples for specific deterrence and general deterrence are defined by these two median values: 1 inspection and 0.583 inspections per (other) facility. The two enforcement-related subsamples for specific deterrence and general deterrence are defined by these two median values: 0 actions and 0 actions per (other) facility. For the second set of behavioral forms, the two inspection-related subsamples for specific deterrence and general deterrence are defined by these two median values: 1 inspection and 0.575 inspections per (other) facility. The two enforcement-related subsamples for specific deterrence and general deterrence are defined by these two median values: 0 actions and 0 actions per (other) facility. For the third set of behavioral forms, the two inspection-related subsamples for specific deterrence and general deterrence are defined by these two median values: 1 inspection and 0.587 inspections per (other) facility. The two enforcement-related subsamples for specific deterrence and general deterrence are defined by these two median values: 0 actions and 0.012 actions per (other) facility.

When exploring the effects of injunctions and SEPs, we examine only behavioral forms measured in each of three calendar years. All of the relevant subsamples are defined by the median value of 0, except SEP-related specific deterrence, which is not relevant to our sample since none of the survey respondents ever received a SEP sanction during the relevant sample period.

These identified median values do not evenly split the sample into two subsamples in most cases, especially when the median value equals 0 (in the case of most enforcement) or 1 (in the case of inspection-based specific deterrence), which limits our ability to test whether greater deterrence leads to better behavior in certain cases.

3. In the case of environmental employees in absolute terms, the Wilcoxon two-sample test reveals statistical significance, but the median two-sample test does not ($p=0.144$). In the case of environmental employees relative to overall employees, the Wilcoxon two-sample test does not reveal statistical significance, but the median two-sample test does ($p=0.055$). In the cases of environmental employees relative to overall employees and audit count, the ranking of the subsample mean values does not coincide with the ranking of subsamples based on the Wilcoxon two-sample test and median two-sample test, which employ scoring methods for assessing whether the distribution of one subsample is located at higher levels of the variable in question than the distribution of the other subsample. For example, the above-median-deterrence subsample mean value of 4.9 is smaller than the below-median-deterrence subsample mean value of 6.1, yet the Wilcoxon two-sample test statistic of 1.4

indicates that the above-median-deterrence subsample is distributed at greater audit counts than is the below-median-deterrence subsample.

4. All of the p-values associated with the correlations exceed 0.10.

5. In absolute terms, this relationship is nearly significant (p = 0.122).

6. In the case of environmental employees relative to overall employees, the Wilcoxon two-sample test does not reveal statistical significance, but the median two-sample test does (p = 0.082). In the cases of environmental employees relative to overall employees, training attendance, and external consultant expenditures relative to overall employees, the ranking of the subsample mean values does not coincide with the ranking of subsamples based on the Wilcoxon two-sample test and median two-sample test, which employ scoring methods for assessing whether the distribution of one subsample is located at higher levels of the variable in question than the distribution of the other subsample.

7. In the cases of environmental employees in absolute terms and environmental employees' education, the Wilcoxon two-sample test not reveal statistical significance, but the median two-sample test does (p = 0.04 and p = 0.08, respectively). In the case of environmental employees relative to overall employees, the ranking of the subsample mean values does not coincide with the ranking of subsamples based on the Wilcoxon two-sample test and median two-sample test, which employ scoring methods for assessing whether the distribution of one subsample is located at higher levels of the variable in question than the distribution of the other subsample.

8. The ranking of the subsample mean values does not coincide with the ranking of subsamples based on the Wilcoxon two-sample test and median two-sample test.

9. In the cases of environmental employees in absolute terms and audit count, the Wilcoxon two-sample tests reveal statistical significance, but the median two-sample tests do not (p = 0.15 and p = 0.14, respectively).

10. For details about the fixed-effects estimator, the interested reader is encouraged to read Wooldridge (2002). The eager reader is encouraged to read Hsiao (1986). To the interested reader, we acknowledge that these two books indicate that the random effects estimator is also appropriate for exploiting and accommodating the panel data structure; use of a random-effects estimator generates conclusions identical to those reported from the use of a fixed-effects estimator.

11. The interested reader is referred to Greene (2003) for an in-depth description of the probit estimator. The eager reader is encouraged to read Maddala (1983). To the interested reader, we acknowledge that these two books indicate that the logit estimator is also appropriate for exploring qualitative dependent variables representing two categories. Generally, these two estimators generate similar results, so the choice between the two need not be important.

12. The interested reader is referred to Wooldridge (2002) for an in-depth description of the Poisson estimator. The eager reader is encouraged to read Cameron and Trivedi (1998). To the interested reader, we acknowledge that these two books indicate that the negative binomial estimator is also appropriate for estimating count-data-dependent variables; preliminary analysis reveals that use of

this alternative estimator generates conclusions identical to those generated by a Poisson estimator.

13. To the interested reader, these unknown aspects represent the error terms within the multivariate regression analysis. For details about the random-effects Poisson estimator, the interested reader is encouraged to read Wooldridge (2002). The eager reader is encouraged to read Cameron and Trivedi (1998). To the interested reader, we acknowledge that the latter book indicates that the fixed-effects Poisson estimator is also appropriate for estimating panel-count-data-dependent variables. However, application of the fixed-effects Poisson estimator requires the elimination of any facility that does not vary its audit count over the three-year sample period, which is true for a meaningful portion of our sample. Consequently, we avoid the use of this estimator.

14. Due to lack of variation, we cannot assess specific deterrence from formal nonsanctions.

15. Due to limited variation in the dependent variable in combination with certain deterrence factors, we cannot assess specific deterrence from formal nonsanctions in both cases and cannot assess general deterrence from fines in the case of TSS treatment presence.

16. Due to limited variation in the dependent variable in combination with certain deterrence factors, we cannot assess specific deterrence from informal actions and formal nonsanctions in both cases and cannot assess specific deterrence from federal inspections in the case of BOD treatment extent. Clearly, the multivariate regression analysis is not able to identify the links between government interventions that are used infrequently (for example, formal nonsanctions) and certain qualitative behavioral decisions (for example, treatment technology: presence versus absence).

17. Due to limited variation, we cannot assess specific deterrence from civil formal nonsanctions.

18. Due to limited variation in the dependent variable in combination with certain deterrence factors, we cannot assess specific deterrence from formal nonsanctions.

19. The associated statistical significance of the latter set of results is, however, sensitive to the manner of measuring injunctions and SEPs. If measured by the count of injunctions or SEPs, the estimated effect is statistically significant; if measured by the value of injunctions or SEPs, the estimated effect is significant only at levels far above 10 percent.

20. Due to limited variation in the dependent variables in combination with certain deterrence factors, we cannot assess specific deterrence from formal nonsanctions in both dimensions of audit quality and specific deterrence from informal enforcement in the case of audit classification protocol.

21. Due to limited variation in the dependent variable in combination with certain deterrence factors, the multivariate analysis is not able to assess specific deterrence from informal enforcement actions and formal nonsanctions and general deterrence from fines.

22. Due to limited variation in the dependent variable in combination with certain deterrence factors, we cannot assess specific deterrence from both federal

inspections and formal nonsanctions and general deterrence from civil formal nonsanctions.

Chapter Nine

1. The survey responses we assess here were provided by all of the respondents to our survey, rather than only those 97 facilities for which the Permit Compliance System database provides information on discharge limits and measurements. The analysis in this section is derived from the description of the perceptions of the effectiveness of government interventions on environmental performance provided in Glicksman and Earnhart (2007a).

2. We phrased the survey questions relating to the effects of government interventions on environmental performance in terms of whether the respondents thought that the particular interventions concerned "are effective ways for inducing individual chemical facilities to comply with permitted water discharge limits." This phrasing captures both specific and general deterrence. The questions refer to "individual" facilities, which carries the flavor of specific deterrence, but they refer to individual "facilities," which might be taken as a reference to the effect of fines at one facility on the industry as a whole.

3. We also assess whether there exist any regional patterns to the responses to the question on the efficacy of inspections on performance. Facilities located in southern states (especially Texas) are somewhat more likely than those in other regions of the country to say that inspections are effective. Facilities in midwestern states such as Iowa and Illinois are more likely to say that inspections are not effective.

4. According to Hodas (1995, p. 1614), "For all practical purposes, administrative enforcement activity is not subject to public scrutiny; thus there is no independent public check on state and federal enforcement practices. In comparison, the judicial actions are open to the public."

5. For this statistical assessment, we denote each classified respondent with one of three values $\{-1, 0, +1\}$. We classify each respondent as granting a higher level of effectiveness to fines $(+1)$, a higher level to injunctions (-1), or an identical level to both (0). Then we assess whether the mean response across all of the respondents is positive, that is, fines are more effective than injunctions. The mean of 0.1141 is significantly greater than 0, with a p-value of 0.029.

6. For this statistical assessment, we denote each classified respondent with one of three values $\{-1, 0, +1\}$. We classify each respondent as granting a higher level of effectiveness to fines $(+1)$, a higher level to SEPs (-1), or an identical level to both (0). Then we assess whether the mean response across all of the respondents is positive; that is, fines are more effective than SEPs. The mean of 0.0615 is insignificantly greater than 0, with a p-value of 0.217.

7. The interested reader is encouraged to read Wooldridge (2002) for more details on the random-effects estimator.

8. As demonstrated by Hausman test statistics, in general, the random-effects estimates appear inconsistent; that is, they are not likely to reflect the "true" relationship between environmental performance and the multiple explanatory factors.

The interested reader is encouraged to read Wooldridge (2002) for more details on this concern.

9. For the interested reader, we state that by omitting environmental behavioral measures from the explanatory factor set, the estimated relationship between performance and interventions involves "omitted variable bias." Given the purpose of the described analysis, we are not concerned by this bias since we are not claiming to interpret only the direct effect of interventions on performance. On the contrary, the notion of "omitted variable bias" helps us to interpret the reflection of the indirect effects of interventions on performance since the bias exists only when interventions and behavior are correlated, which is established in general in Chapter 8.

10. Hausman test statistics reveal that the random-effects estimates appear consistent in the case of TSS quantity-based absolute discharges and relative discharges. Respectively, the test statistics are 15.85 and 18.82, with p-values of 0.257 and 0.129, given 13 degrees of freedom. In all other cases, the Hausman test statistics fail to indicate that the random-effects estimates appear consistent.

11. For quantity-based, production-adjusted absolute discharges and relative discharges: respectively, $\beta = -16.7, \beta = -0.206; p = 0.000, p = 0.001$. For concentration-based absolute and relative discharges: respectively, $\beta = -6.94, \beta = -0.255; p = 0.074, p = 0.013$.

12. When not adjusted for production, $\beta = -65.0$ and $\beta = -0.353$, respectively, and p=0.018 and p=0.053, respectively.

References

Aden, Jean, Ahn Kyu-Hong, and Michael T. Rock. 1999. "What Is Driving the Pollution Abatement Expenditure Behavior of Manufacturing Plants in Korea?" *World Development* 27 (7): 1203–14.

Anderson, Eric E., and Wayne K. Talley. 1995. "The Oil Spill Size of Tanker and Barge Accidents: Determinants and Policy Implications." *Land Economics* 71 (2): 216–28.

Anton, Wilma Rose Q., George Deltas, and Madhu Khanna. 2003. "Incentives for Environmental Self-Regulation and Implications for Environmental Performance." *Journal of Environmental Economics and Management* 48:632–54.

Arimura, Toshi H., Akira Hibiki, and Hajime Katayama. 2008. "Is a Voluntary Approach an Effective Environmental Policy Instrument? A Case for Environmental Management Systems." *Journal of Environmental Economics and Management* 55:281–95.

Barla, Philippe. 2007. "ISO 14001 Certification and Environmental Performance in Quebec's Pulp and Paper Industry." *Journal of Environmental Economics and Management* 53:291–306.

Berman, Eli, and Linda T. M. Bui. 2001a. "Environmental Regulation and Labor Demand: Evidence from the South Coast Air Basin." *Journal of Public Economics* 79:265–95.

———. 2001b. "Environmental Regulation and Productivity: Evidence from Oil Refineries." *Review of Economics and Statistics* 83 (3): 498–510.

Bluffstone, Randall, and Thomas Sterner. 2006. "Explaining Environmental Management in Central and Eastern Europe." *Comparative Economic Studies* 48:619–40.

Brännlund, Runar, and Karl-Gustaf Löfgren. 1996. "Emission Standards and Stochastic Waste Load." *Land Economics* 72 (2): 218–30.

Cameron, A. Colin, and Pravin K. Trivedi. 1998. *Regression Analysis of Count Data*. Econometric Society Monograph 30. Cambridge: Cambridge University Press.

Christmann, Petra, and Glen Taylor. 2001. "Globalization and the Environment: Determinants of Firm Self-Regulation in China." *Journal of International Business Studies* 32 (3): 439–58.

Cohen, Mark A. 1987. "Optimal Enforcement Strategy to Prevent Oil Spills: An Application of a Principal-Agent Model with Moral Hazard." *Journal of Law and Economics* 30 (1): 23–51.

Collins, Allan, and Richard I. D. Harris. 2002. "Does Plant Ownership Affect the Level of Pollution Abatement Expenditure?" *Land Economics* 78 (2): 171–89.

Cropper, Maureen L., William N. Evans, Stephen J. Berardi, Maria M. Ducla-Soares, and Paul R. Portney. 1992. "The Determinants of Pesticide Regulation: A Statistical Analysis of EPA Decision Making." *Journal of Political Economy* 100 (1): 175–97.

Dana, David A. 1998. "The Uncertain Merits of Environmental Enforcement Reform: The Case of Supplemental Environmental Projects." *Wisconsin Law Review* 1998:1181–21.

Dasgupta, Susmita, Hemamala Hettige, and David Wheeler. 2000. "What Improves Environmental Compliance? Evidence from Mexican Industry." *Journal of Environmental Economics and Management* 39:39–66.

Dasgupta, Susmita, Benoit Laplante, Nlandu Maminig, and Hua Wang. 2001. "Inspections, Pollution Prices, and Environmental Performance: Evidence from China." *Ecological Economics* 36:487–98.

Deily, Mary, and Wayne Gray. 1991. "Enforcement of Pollution Regulation in a Declining Industry." *Journal of Environmental Economics and Management* 21:260–74.

Delmas, Magali D., and Michael W. Toffel. 2008. "Organizational Responses to Environmental Demands: Opening the Black Box." *Strategic Management Journal* 29:1027–55.

DeShazo, J. R., and Andres Lerner. 2004. "The Consequences of Devolution for Standard Setting: An Empirical Analysis of the Clean Water Act." Working paper, University of California, Los Angeles, December 1, 2004.

Duhigg, Charles. 2009. "Clean Water Laws Neglected, at a Cost." *New York Times*, September 13, 2009.

Earnhart, Dietrich. 2004a. "The Effects of Community Characteristics on Polluter Compliance Levels." *Land Economics* 80 (3): 408–32.

———. 2004b. "Panel Data Analysis of Regulatory Factors Shaping Environmental Performance." *Review of Economics and Statistics* 86 (1): 391–401.

———. 2004c. "Regulatory Factors Shaping Environmental Performance at Publicly-Owned Treatment Plants." *Journal of Environmental Economics and Management* 48 (1): 655–81.

Earnhart, Dietrich and Lubomir Lizal. 2006. "Effects of Ownership and Financial Performance on Corporate Environmental Performance," *Journal of Comparative Economics* 34 (1): 111–29.

Eckert, Heather. 2004. "Inspections, Warnings, and Compliance: The Case of Petroleum Storage Regulation." *Journal of Environmental Economics and Management* 47 (2): 232–59.

EPA. *See* U.S. Environmental Protection Agency.

Epple, Dennis, and Michael Visscher. 1984. "Environmental Pollution: Modeling Occurrence, Detection, and Deterrence." *Journal of Law and Economics* 27 (1): 29–60.

Evans, Mary, Lirong Liu, and Sarah Stafford. 2008. "Causes and Consequences of Environmental Auditing: Evidence from Regulated Facilities in Michigan." College of William and Mary Working Paper 78, Williamsburg, VA.

Farber, Daniel A. 1984. "Equitable Discretion, Legal Duties, and Environmental Injunctions." *University of Pittsburgh Law Review* 45:413–545.

Foulon, Jérôme, Paul Lanoie, and Benôit Laplante. 2002. "Incentives for Pollution Control: Regulation or Information." *Journal of Environmental Economics and Management* 44 (1): 169–87.

Frondel, Manuel, Jens Horbach, and Klaus Rennings. 2007. "End-of-Pipe or Cleaner Production Measures? An Empirical Comparison of Abatement Decisions Across OECD Countries." *Business Strategy and the Environment* 16 (8): 571–84.

Gallagher, Lynn M., and Leonard Miller. 1996. *Clean Water Handbook*. Rockville, MD: Government Institutes.

Gangadharan, Lata. 2006. "Environmental Compliance by Firms in the Manufacturing Sector in Mexico." *Ecological Economics* 59:477–86.

GAO. *See* U.S. Government Accountability Office.

Glicksman, Robert, and Dietrich Earnhart. 2007a. "The Comparative Effectiveness of Government Interventions on Environmental Performance in the Chemical Industry." *Stanford Environmental Law Journal* 26:317371.

———. 2007b. "Depiction of the Regulator-Regulated Entity Relationship in the Chemical Industry: Deterrence-Based vs. Cooperative Enforcement." *William and Mary Environmental Law and Policy Review* 31 (3): 603–60.

Gray, Wayne B., and Mary E. Deily. 1996. "Compliance and Enforcement: Air Pollution Regulation in the U.S. Steel Industry." *Journal of Environmental Economics and Management* 31:96–111.

Gray, Wayne B., and Ronald J. Shadbegian. 2005. "When and Why Do Plants Comply? Paper Mills in the 1980s." *Law and Policy* 27 (2): 238–61.

Greene, William. 2003. *Econometrics*. 5th ed. Englewood Cliffs, NJ: Prentice Hall.

Heckman, James. 1979. "Sample Selection Bias as Specification Error." *Econometrica* 47:153–61.

Helland, Eric. 1998a. "The Enforcement of Pollution Control Laws: Inspections, Violations and Self-Reporting." *Review of Economics and Statistics* 80 (1): 141–53.

———. 1998b. "The Revealed Preferences of State EPAs: Stringency, Enforcement, and Substitution." *Journal of Environmental Economics and Management* 35 (3): 242–61.

Henriques, Irene, and Perry Sadorsky. 2006. "The Adoption of Environmental Management Practices in a Transition Economy." *Comparative Economic Studies* 48:641–61.

Hodas, David R. 1995. "Enforcement of Environmental Law in a Triangular Federal System: Can Three Not Be a Crowd When Enforcement Authority Is Shared by the United States, the States, and Their Citizens?" *Maryland Law Review* 54:1552–1657.

Hsiao, Cheng. 1986. *Analysis of Panel Data*. Cambridge: Cambridge University Press.

Hunter, Susan, and Richard Waterman. 1992. "Determining an Agency's Regulatory Style: How Does the EPA Water Office Enforce the Law?" *Western Political Quarterly* 45:403.

Kagan, Robert, Neil Gunningham, and Dorothy Thornton. 2003. "Explaining Corporate Environmental Performance: How Does Regulation Matter?" *Law and Society Review* 37:51–84.

Kerr, Suzi, and Richard G. Newell. 2003. "Policy-Induced Technology Adoption: Evidence from the U.S. Lead Phasedown." *Journal of Industrial Economics* 51 (3): 317–43.

Khanna, Madhu, Patricia Koss, Cody Jones, and David Ervin. 2007. "Motivations for Voluntary Environmental Management." *Policy Studies Journal* 35 (4): 751–72.

Khanna, Madhu, and Wilma Rose Q. Anton. 2002. "Corporate Environmental Management: Regulatory and Market-Based Incentives." *Land Economics* 78 (4): 539–58.

Lanoie, Paul, Mark Thomas, and Joan Fearnley. 1998. "Firms and Responses to Effluent Regulations: Pulp and Paper in Ontario, 1985–1989." *Journal of Regulatory Economics* 13:103–20.

Laplante, Benoît, and Paul Rilstone. 1996. "Environmental Inspections and Emissions of the Pulp and Paper Industry in Quebec." *Journal of Environmental Economics and Management* 31 (1): 19–36.

Maddala, G.S. 1983. *Limited-Dependent and Qualitative Variables in Econometrics*. Cambridge: Cambridge University Press.

Magat, Wesley A., Alan J. Krupnick, and Winston Harrington. 1986. *Rules in the Making: A Statistical Analysis of Regulatory Agency Behavior*. Washington, DC: Resources for the Future.

Magat, Wesley A., and W. Kip Viscusi. 1990. "Effectiveness of the EPA's Regulatory Enforcement: The Case of Industrial Effluent Standards." *Journal of Law and Economics* 33 (2): 331–60.

Markell, David. 2007. "Is There a Possible Role for Regulatory Enforcement in the Effort to Value, Protect, and Restore Ecosystem Services?" *Journal of Land Use and Environmental Law* 22:549–98.

Maynard, Leigh J., and James S. Shortle. 2001. "Determinants of Cleaner Technology Investments in the U.S. Bleached Kraft Pulp Industry." *Land Economics* 77 (4): 561–76.

Mickwitz, Per. 2003. "Is It as Bad as It Sounds or as Good as It Looks? Experiences of Finnish Water Discharge Limits." *Ecological Economics* 45:237–54.

Mori, Yasuhumi, and Eric W. Welch. 2008. "The ISO 14001 Environmental Management Standard in Japan: Results from a National Survey of Facilities in Four Industries." *Journal of Environmental Planning and Management* 51 (3): 421–45.

Nadeau, Louis W. 1997. "EPA Effectiveness at Reducing the Duration of Plant-Level Noncompliance." *Journal of Environmental Economics and Management* 34 (1): 54–78.

Nakamura, Masao, Takuya Takahashi, and Ilan Vertinsky. 2000. "Why Japanese Firms Choose to Certify: A Study of Managerial Responses to Environmental Issues." *Journal of Environmental Economics and Management* 42:23–52.

New York Times. 2009. "Clean Water: Still Elusive." October 22, 2009.

Porter, Michael, and Claas van der Linde. 1995. "Toward a New Conception of the Environment-Competitiveness Relationship." *Journal of Economic Perspectives* 9 (4): 97–118.

Rechtschaffen, Clifford. 2004. "Enforcing the Clean Water Act in the Twenty-first Century: Harnessing the Power of the Public Spotlight." *Alabama Law Review* 55 (56): 775–814.

Rechtschaffen, Clifford, and David Markell. 2003. *Reinventing Environmental Enforcement and the State/Federal Relationship.* Washington, DC: Environmental Law Institute.

Rousseau, Jean-Jacques. 1755. *The Social Contract and Discourse on Political Economy.* Reprinted in *The Social Contract and Discourses,* trans. G.D.H. Cole. New York: E.P Dutton, 1923.

Shapiro, Sidney A., and Robert L. Glicksman. 2000. "Goals, Instruments, and Environmental Policy Choice." *Duke Environmental Law and Policy Forum* 10:297–325.

Shimshack, Jay P., and Michael B. Ward. 2005. "Regulator Reputation, Enforcement, and Environmental Compliance." *Journal of Environmental Economics and Management* 50 (3): 519–40.

Silverman, Samuel L. 1990. "Federal Enforcement of Environmental Laws." *Massachusetts Law Review* 75:95–98.

Smith, Henry E. 2000. "Ambiguous Quality Changes from Taxes and Legal Rules." *University of Chicago Law Review* 67:647–723.

Snyder, Lori D., Nolan H. Miller, and Robert N. Stavins. 2003. "The Effects of Environmental Regulation on Technology Diffusion: The Case of Chlorine Manufacturing." *American Economic Review* 93 (2): 431–35.

Stafford, Sarah L. 2002. "The Effect of Punishment on Firm Compliance with Hazardous Waste Regulations." *Journal of Environmental Economics and Management* 44 (2): 290–308.

U.S. Environmental Protection Agency. 1986. "Evaluation Area Descriptive Guidance for New England Water (NPDES) Programs, App. 2 to EPA Office of Enforcement and Compliance Monitoring, Policy Framework for State/EPA Enforcement Agreements." U.S. Environmental Protection Agency, Washington, DC, August.

———. 1990. "A Primer on the Office of Water Enforcement and Permits and Its Programs." Office of Water, U.S. Environmental Protection Agency, Washington, DC, March.

———. 1994. "NPDES Compliance Inspection Manual." Office of Enforcement and Compliance Assurance, U.S. Environmental Protection Agency, Washington, DC, September.

———. 1995. "Interim Clean Water Act Settlement Policy." U.S. Environmental Protection Agency, Washington, DC.

———. 1996. "U.S. EPA NPDES Permit Writer's Manual." Office of Water, U.S. Environmental Protection Agency, EPA-833-B-96-003, December.

———. 1997. *Chemical Industry National Environmental Baseline Report 1990–1994.* Office of Enforcement and Compliance Assurance, U.S. Environmental Protection Agency, EPA 305-R-96-002, October.

————. 1998. "EPA Supplemental Environmental Projects Policy." Office of Enforcement and Compliance Assurance, U.S. Environmental Protection Agency, *Federal Register* 63 (May 5): 24, 796–804.

————. 1999. *EPA/CMA Root Cause Analysis Pilot Project: An Industry Survey.* Environmental Protection Agency, EPA-305-R-99-001, May.

————. 2001. "Beyond Compliance: Supplemental Environmental Projects." Office of Enforcement and Compliance Assurance, U.S. Environmental Protection Agency, Washington, DC, January.

————. 2009. *Clean Water Act Enforcement Action Plan.* Office of Enforcement and Compliance Assurance, U.S. Environmental Protection Agency, October 15.

U.S. Government Accountability Office. 1996. "Water Pollution: Differences Among the States in Issuing Permits Limiting the Discharge of Pollutants." United States General Accounting Office, GAO/RCED-96-42, January.

————. 2009. Testimony Before the Committee on Transportation and Infrastructure, U.S. House of Representatives, *Clean Water Act: Longstanding Issues Impact EPA's and States' Enforcement Efforts.* Statement of Ann K. Mittal, Director, Natural Resources and Enforcement Team, United States Government Accountability Office, GAO-10-165T, October 15.

Viladrich-Grau, Montserrat, and Theodore Groves. 1997. "The Oil Spill Process: The Effect of Coast Guard Monitoring on Oil Spills." *Environmental and Resource Economics* 10 (4): 315–39.

Wang, Hua. 2002. "Pollution Regulation and Abatement Efforts: China." *Ecological Economics* 41:85–94.

Wang, Hua, Nlandu Mamingi, Benoit Laplante, and Susmita Dasgupta. 2003. "Incomplete Enforcement of Pollution Regulation: Bargaining Power of Chinese Factories." *Environmental and Resource Economics* 24:245–62.

Wang, Hua, and David Wheeler. 2005. "Financial Incentives and Endogenous Enforcement in China's Pollution Levy System." *Journal of Environmental Economics and Management* 49:174–96.

Wooldridge, Jeffrey. 2002. *Econometric Analysis of Cross Section and Panel Data.* Cambridge, MA: MIT Press.

Zinn, Matthew D. 2002. "Policing Environmental Regulatory Enforcement: Cooperation, Capture, and Citizen Suits." *Stanford Environmental Law Journal* 21:81–174.

Index

American Chemical Council, 17
Audits, environmental: quality (classification protocol, team composition), 87, 117, 123, 131, 139, 145, 180–181, 252; quantity, 86–87, 117, 122–123, 131, 139, 144–145, 180–181, 252

Biological oxygen demand (BOD): definition of, 47; importance of, 47–48

Chemical industry: effluent limitation guidelines for different sectors, 38; importance of discharges, 8–9; survey of, 16–21
Clean Water Act: Enforcement Action Plan, 3–4; designated use, 39–42; dredge and fill permit program, 6; effluent limitation guidelines, 6, 25, 36–40, 42–45, 48, 72; goal of eliminating discharge of pollutants, 1, 37–39, 289; fishable, swimmable waters, 6, 33; performance standards, 37, 43; phased system of discharge reduction, 37–38; point sources, regulation of existing sources, 38, 40; point sources, regulation of new sources, 38, 40, 45; reporting requirements, 7, 34; technology-based pollution controls, 36–38, 40; total maximum daily loads (TMDLs), 41–42; violations, in general, 2, 4; water-quality criteria, 41; water-quality standards, 7, 36, 40–43, 56, 288. See also Effluent limits; Surface water quality

Communication of environmental performance goals and targets: to employees, 88, 118, 125, 128, 133, 146, 253; to external parties, 88, 118, 125, 128, 133, 146, 253
Compliance: discharge compliance status, 155–161; economic benefit of noncompliance, 198; extent of compliance, 155–161; noncompliance, 2–3, 17, 30–31, 75, 87, 118, 124, 128, 133, 139–140, 145–46, 157, 163, 193, 196, 198–199, 205, 211, 214–216, 253, 258, 277, 285–286, 289, 295. See also Discharges
Conclusions: policy implications of, 287–292; summary of, 283–287

Deterrence: general deterrence, definition of, 219–221; specific deterrence, definition of, 218–219
Discharge limits. See Effluent limits
Discharges: absolute discharges, assessment of, 155–162; chemical industry, importance of, 8–9; discharge ratios, assessment of, 155–162; relative discharges, assessment of, 155–162

Economic benefit of noncompliance, 198
Effluent limits: best professional judgment (BPJ), 44, 56; binding limits, definition of, 163; concentration limits, definition of, 45–46; concentration-equivalent limits, construction of, 55–56; effective limits, definition of, 162–163; imposition

Policy Research Institute Survey
  Research Center, 19, 20
Pollutants: conventional pollutants,
  17, 37–38, 47–48; toxic pollutants,
  37–38, 48, 205
Previous studies, context of, 25–32
Priority industrial sectors, 17

Reporting: monthly discharge
  monitoring reports, 194; strategic
  nonreporting of discharges, 156.
  See also Clean Water Act: reporting
  requirements
Research questions, summary of, 24–25
Responsible Care program, 17

Sample of regulated facilities: in general,
  16–22; sample selection bias, 20–21.
  See also Survey of chemical industry
Self-audits. *See* Audits, environmental
Significant noncompliance, rates of, 3
State regulators: cooperative
  relationships with, 256; enforcement

budgets, 2; joint administration with
  EPA, 34–35
Surface water quality, 3–4, 7, 11, 13, 18,
  36, 41, 43, 46–47, 199. *See also* Clean
  Water Act: water-quality standards
Survey of chemical industry, 16–21. *See
  also* Chemical industry; Sample of
  regulated facilities

Total suspended solids (TSS): definition
  of, 47; importance of, 47–48
Treatment technologies: effects of
  government interventions on, 251;
  effect of limits on, 115–116, 121–122,
  127, 130; effects on environmental
  performance, 178–180; treatment of
  specific pollutants, 47–48; types of,
  86; upgrades to, 86

Wastewater discharge limits. *See* Effluent
  limits
Wastewater discharges. *See* Discharges
Water quality. *See* Surface water quality

The authorized representative in the EU for product safety and compliance is:
Mare Nostrum Group
B.V Doelen 72
4831 GR Breda
The Netherlands

www.ingramcontent.com/pod-product-compliance
Lightning Source LLC
Chambersburg PA
CBHW021809270326
41932CB00007B/110